VISUALIZING CHEMISTRY

CHEMICAL POTENTIAL ENERGY & CHEMICAL SYSTEMS

JERRY P. SUITS

University of Northern Colorado

Nitroglycerin

$C_3H_5(ONO_2)_3$

E(act) = 82 kJ/mol

ΔH(comb) = -1548 kJ/mol

$3\ CO_2\ (g) + 5/2\ H_2O\ (g) + 3/2\ N_2\ (g)$

Progress of Reaction

Cover images © Shutterstock.com

www.kendallhunt.com
Send all inquiries to:
4050 Westmark Drive
Dubuque, IA 52004-1840

ISBN 978-1-4652-8830-1

Printed in the United States of America

Brief Contents

Unit # (and corresponding chapter in published general chemistry texts)

	Preface	**vii**
Unit 1	**Visualizing Chemical Potential Energy** (no equivalent chapter in any text)	**1**
Unit 2	**Visualizing Chemical Reaction Rates** (Chapter 12, 13, or 14 in most texts)	**23**
Unit 3	**What Is a Chemical System?** (no equivalent chapter in any text)	**35**
Unit 4	**Equilibrium in Chemical Systems** (Chapter 13, 14, or 15) . . .	**41**
Unit 5	**Acid-Base Aqueous Systems** (Chapters 14, 15, 16, or 17—any two of these)	**53**
Unit 6	**Solubility of Aqueous Salts** (Chapter 16 or 17)	**69**
Unit 7	**Thermodynamic Systems** (Chapter 17 or 18)	**81**

Contents

Preface vii

Unit 1 Visualizing Chemical Potential Energy 1

1.1 Introduction 1
1.2 Law of Conservation of Energy 4
1.3 Chemical Potential Energy Diagrams 6
1.4 Activation Energy 7
1.5 Chemical Kinetics 8
1.6 Chemical Potential Energy of Explosives 10
1.7 Oxidation and Combustion 13
1.8 Electronegativity Values for Nonmetal Elements 13
1.9 Fuels and Chemical Potential Energy 13
1.10 Chemical Bonding of the Common Nonmetal Elements 16
1.11 Does Breaking Bonds Produce Energy? 17
1.12 Summary 19
1.13 Technical References 19
1.14 Questions to Ponder 20

Unit 2 Visualizing Chemical Reaction Rates 23

2.1 Introduction 23
2.2 Order of a Reaction 24
2.3 Temperature, Kinetic Energy, and E_{act} 28
2.4 Colliding Molecules and Order of Reaction 29
2.5 Effect of a Catalyst on E_{act} 30
2.6 How Can We Compare Various Reaction Rates? 31
2.7 Summary 32
2.8 Technical References 33
2.9 Questions to Ponder 34

Unit 3 What Is a Chemical System? 35

3.1 Introduction 35
3.2 What Are Interacting Chemical Species? 36
3.3 Dynamic Reactant/Product Relationships 36
3.4 Physical/Chemical Condition 39
3.5 Predictable Manner of Chemical Behavior 40
3.6 Changes in Reaction Parameters (Conditions) 40
3.7 What Exactly Is a Chemical System? 40

Unit 4 Equilibrium in Chemical Systems 41

4.1 Introduction 41
4.2 Reversible Reactions and Equilibrium 42
4.3 Reversible Reactions and the Meaning of K_c 44
4.4 Reversible Reactions and Moderate Values of K_c 44
4.5 Reversible Reactions and Very Large Values of K_c 46
4.6 Reversible Reactions and Very Small Values of K_c 48
4.7 Reversible Reactions and Non-equilibrium 49
4.8 Summary 51
4.9 Technical References 52

Unit 5 Acid-Base Aqueous Systems 53

5.1 Introduction 53
5.2 Acids and Degree of Acidity 55
5.3 Bases and Their Degree of Ionization 57
5.4 Visualizing the pH Scale 58
5.5 Monoprotic Acid Distribution Systems across pH 61
5.6 Diprotic Acid Distribution Systems across pH 64
5.7 Summary 67
5.8 Technical References 67

Unit 6 Solubility of Aqueous Salts 69

6.1 Introduction 69
6.2 K_{sp} for a 1:1 Cation:Anion System 71
6.3 K_{sp} for a 3:2 Cation:Anion System 72
6.4 Q_{sp}: Non-Equilibrium Solubility of a Slightly Soluble Salt 74
6.5 Chemical Potential Energy: K_{sp} *vs.* Q_{sp} 77
6.6 Predicting Whether Precipitation Will Occur 78
6.7 Summary 79
6.8 Technical References 79

Unit 7 Thermodynamic Systems 81

7.1 Introduction 81
7.2 Microstates and Entropy 82
7.3 When Is an Exothermic Reaction Spontaneous? 85
7.4 Can an Endothermic Process Produce a Spontaneous Reaction? 88
7.5 When Is a Gas-Phase Reaction Spontaneous? 89
7.6 What Are the Four Types of Thermodynamic Reactions? 92
7.7 Summary 93
7.8 Technical References 94

Preface

This book is intended to supplement *second-semester general chemist*ry for science majors. It shows students how the two fundamental concepts of this course (see Units #1 and #3) relate to the topics normally taught in this course. Furthermore, these two concepts also provide the *conceptual chemistry* needed for students to understand all subsequent chemistry courses (e.g., organic chemistry, biochemistry, and physical chemistry, and molecular biology). Conceptual chemistry for the first semester course has already been established, and it appears in all modern general chemistry textbooks (from 1995 to the present).

To the Student:

The second semester of general chemistry is a very mathematical intensive course; however, even the ACS standardized final examination in this course only contains about 25% problems requiring the use of your calculator. Well, what about the other 75% of questions? They require that you understand concepts and mathematical trends. What are the concepts that underlie these topics? They are . . . *chemical potential energy* and *chemical systems*. Does your textbook even mention these two concepts? The answer is . . . probably not because they seem so obvious to most chemists. Practicing chemists use these two concepts on a daily basis to understand their research or applications. However, chemists tend to think of these concepts as visual images in the mind. Well, why don't they share these important concepts with their students? The reason is that it is difficult to transfer a visual image from one mind to another. If you are a Star Trek fan—Mr. Spock could do a "mind meld" and read what was on another person's mind. However, most students do not possess this "talent" and your chemistry instructor may get upset if you try to read his/her mind in this manner. Thus, we have the rationale for this book: The goal is to show you visualizations that are in the chemists' mindS when they are thinking chemistry.

How should you use this book to understand chemistry? You should read a particular unit before it is covered by your instructor in lecture. This book is written in a conversational tone… one that the author hopes you can read and comprehend what you are reading. Yes, the style is very informal but it is supposed to be just like a chemistry tutor would be talking to you about chemistry topics. When should you read your main textbook? The answer is when you want to see *worked-out problems* that show how to do the mathematical operations in a particular chapter. Should you sit down and read an entire chapter in your main textbook? No! The book's formal style is not conducive to this approach. That is, it is far better just to read one section at a time, then work problems that are covered—just from that one section. This is called "divide and conquer"… *Poco a poco*—little by little. Otherwise, reading the entire chapter in one sitting is like taking three or four wooden baseball bats and trying to break them apart in the middle of the bats. PS: If you try to do this… do not use your head—that could give you a real headache!

Overall, good luck and may the *Chemical Force* (which is the gradient of the chemical potential energy) be with you!

Cheers,
Jerry P. Suits, PhD
December 2015

UNIT 1
Visualizing Chemical Potential Energy

WHY STUDY THIS UNIT?

The goal of this unit is to help you begin to understand "chemical potential energy" as a fundamental concept that is used in all chemistry courses—from second semester general chemistry to organic chemistry to biochemistry and beyond.

Sec 1.1 Introduction

Understanding chemical reactions involves much more than just "balancing chemical equations" because they depend on how stable the chemical species are in both the reactant state and the product state. First, what are chemical species? Molecules, atoms, and ions are all chemical species. For example, when sodium metal (which is composed of sodium atoms) is dropped into water, a chemical reaction proceeds to produce sodium hydroxide and hydrogen gas.

EQ 1.1: $2\ Na\ (s) + 2\ H_2O\ (l) \rightarrow 2\ NaOH\ (aq) + H_2\ (g)$

Metal — liquid water — aqueous soln — explosive gas

However, when gold is dropped into water, there is no reaction. That is, you could wait over 200 years for the reaction to occur, but after that amount of time you would only find gold metal and water (assuming that they are placed in a sealed container to avoid evaporation of water)*.

EQ 1.2: $Au\ (s) + H_2O\ (l) \rightarrow$ No Rxn

Metal — (Gold can be fully recovered)

Thus, sodium metal is a more reactive chemical species than is gold. This means that sodium has a much higher chemical potential energy than does gold.

In the sodium + water example, sodium is a very reactive metal while water is not particularly reactive with most metals. You may recall that this information can be predicted from the "metal activity series" (see Table 1.1). This series is usually shown as a vertical list with the most reactive metals at the top and the least reactive metals at the bottom. Another example that uses this information is when a

* In 2007 a treasure of gold and silver coins was salvaged from a sunken Spanish galleon sunk by British warships in 1804. The silver coins were tarnished but the gold coins only needed to be "dusted off." The treasure was worth about $500M.

more reactive metal, say zinc, is placed in a solution containing copper ions (usually Cu^{2+}, which is a blue-colored solution), then a chemical reaction occurs:

EQ 1.3: $Zn\ (s) + Cu(NO_3)_2\ (aq) \rightarrow Zn(NO_3)_2\ (aq) + Cu\ (s)$

Metal — blue solution — colorless soln — copper-colored metal

On the other hand, when copper metal is placed in a zinc ion, Zn^{2+}, solution, no reaction occurs:

EQ 1.4: $Cu\ (s) + Zn(NO_3)_2\ (aq) \rightarrow$ No Rxn

Metal — colorless soln — (copper-colored metal is fully recovered)

Thus, knowing how to balance a chemical equation does not tell you whether or not the reaction actually occurs. Rather, knowing the chemical potential energies of the reactants (and products) is the key to predicting whether or not a chemical reaction should occur. Another factor in understanding this concept is that determining the chemical reactivity/stability of one chemical species is relative to that of another species. For example, iron undergoes a corrosion reaction to form rust*:

EQ 1.5: $Fe\ (s) + O_2\ (g) + H_2O\ (l) \rightarrow Fe_2O_3 \bullet n\ H_2O\ (s)$

(Rust is hydrated iron (III) oxide)

Water, especially salt water, is needed for this reaction to occur. Your grandpa's old car that is sitting in his backyard may be mostly rust-colored. However, there is not much evidence of rusting in the deserts (e.g., Death Valley, CA) due to very little amounts of water and water vapor (humidity). Grandpa's car there could be shiny and, with a little washing and polishing, it could look like new. Fortunately, there is a way to prevent iron from rusting even in humid climates. If iron is treated with a thin layer of zinc, then the iron will not rust. This is because zinc is a more active metal than is iron (see Table 1.1). In other words, zinc has a greater chemical potential energy than iron, and the following reaction occurs:

EQ 1.6: $2\ Zn\ (s) + O_2\ (g) \rightarrow 2\ ZnO\ (s)$

Again, chemical potential energy of a chemical species is *relative* to other chemical species in a chemical reaction (see Table 1.1; Fe vs Zn).

Chemists, of course, understand chemical potential energies, and most of them form visual images in their minds[1], which help them to visualize this fundamental concept. Well, what are you to do to try to understand it? How about if we learn to draw circles of different sizes to represent different chemical potential energies? Thus, a larger circle represents a higher chemical PE while a smaller circle shows a lower chemical PE. A larger chemical potential energy means that that chemical species tends to be more reactive (i.e., less stable). Likewise, a lower chemical PE means that the species is more stable (i.e., less reactive). Let's look again at the zinc metal and copper (II) ion reaction. Zinc metal has a larger chemical PE than does copper metal.

---

* This reaction that forms rust is actually more complex than this equation shows. Rust is a hydrated salt of iron (III) oxide.

TABLE 1.1 Activity Series of Metals and Chemical Potential Energies

Metal element	Metal atom (solid)	Metal cation (aq)	Oxidation ½ Rxn	Chem Rxns *
Lithium			$Li (s) \rightarrow Li^{+} (aq) + e-$	$+ H^{+} (aq) \rightarrow H_2 (g)$ $+ H_2O (l) \rightarrow H_2 (g)$
Potassium			$K (s) \rightarrow K^{+} (aq) + e-$	$+ H^{+} (aq) \rightarrow H_2 (g)$ $+ H_2O (l) \rightarrow H_2 (g)$
Barium			$Ba(s) \rightarrow Ba^{2+} (aq) + 2e-$	$+ H^{+} (aq) \rightarrow H_2 (g)$ $+ H_2O (l) \rightarrow H_2 (g)$
Calcium			$Ca(s) \rightarrow Ca^{2+} (aq) + 2e-$	$+ H^{+} (aq) \rightarrow H_2 (g)$ $+ H_2O (l) \rightarrow H_2 (g)$
Sodium			$Na (s) \rightarrow Na^{+} (aq) + e-$	$+ H^{+} (aq) \rightarrow H_2 (g)$ $+ H_2O (l) \rightarrow H_2 (g)$
Magnesium			$Mg (s) \rightarrow Mg^{2+} (aq) + 2e-$	$+ H^{+} (aq) \rightarrow H_2 (g)$ $+ H_2O (g) \rightarrow H_2 (g)$
Aluminum			$Al (s) \rightarrow Al^{3+} (aq) + 3e-$	$+ H^{+} (aq) \rightarrow H_2 (g)$ $+ H_2O (g) \rightarrow H_2 (g)$
Manganese			$Mn (s) \rightarrow Mn^{2+} (aq) + 2e-$	$+ H^{+} (aq) \rightarrow H_2 (g)$ $+ H_2O (g) \rightarrow H_2 (g)$
Zinc			$Zn (s) \rightarrow Zn^{2+} (aq) + 2e-$	$+ H^{+} (aq) \rightarrow H_2 (g)$ $+ H_2O (g) \rightarrow H_2 (g)$
Chromium			$Cr(s) \rightarrow Cr^{3+} (aq) + 3e-$	$+ H^{+} (aq) \rightarrow H_2 (g)$ $+ H_2O (g) \rightarrow H_2 (g)$
Iron			$Fe (s) \rightarrow Fe^{2+} (aq) + 2e-$	$+ H^{+} (aq) \rightarrow H_2 (g)$ $+ H_2O (g) \rightarrow H_2 (g)$
Cobalt			$Co (s) \rightarrow Co^{2+} (aq) + 2e-$	$+ H^{+} (aq) \rightarrow H_2 (g)$
Nickel			$Ni (s) \rightarrow Ni^{2+} (aq) + 2e-$	$+ H^{+} (aq) \rightarrow H_2 (g)$
Tin			$Sn (s) \rightarrow Sn^{2+} (aq) + 2e-$	$+ H^{+} (aq) \rightarrow H_2 (g)$
Lead			$Li (s) \rightarrow Li^{+} (aq) + e-$	$+ H^{+} (aq) \rightarrow H_2 (g)$
HYDROGEN **			$H_2 (g) \rightarrow 2 H^{+} + 2 e-$	Standard
Copper			$Cu (s) \rightarrow Cu^{2+} (aq) + 2e-$	$+ H^{+} (aq) \rightarrow$ NoRxn
Silver			$Ag (s) \rightarrow Ag^{+} (aq) + e-$	$+ H^{+} (aq) \rightarrow$ NoRxn
Mercury	Hg (l)		$Hg (l) \rightarrow Hg^{2+} (aq) + 2e-$	$+ H^{+} (aq) \rightarrow$ NoRxn
Platinum			$Pt (s) \rightarrow Pt^{2+} (aq) + 2e-$	$+ H^{+} (aq) \rightarrow$ NoRxn
Gold			$Au (s) \rightarrow Au^{3+} + 3 e-$	$+ H^{+} (aq) \rightarrow$ NoRxn

* Chem Rxns: H^{+} (aq) = strong acid, e.g., HCl (aq); H_2O (g) = steam

**HYDROGEN is the "*balancing point*" where chemists assume that the hydrogen reaction is at "true thermodynamic equilibrium," which will be explained in Unit 7 Thermodynamics. Note that the circles representing chemical potential energy for reactant and product are equal areas, i.e., $PE(H_2) = PE(H^{+})$.

EQ 1.3: $Zn\ (s) + Cu(NO_3)_2\ (aq) \rightarrow Zn(NO_3)_2\ (aq) + Cu\ (s)$

Net ionic EQ: $Zn\ (s) + Cu^{2+}\ (aq) \rightarrow Zn^{2+}\ (aq) + Cu$

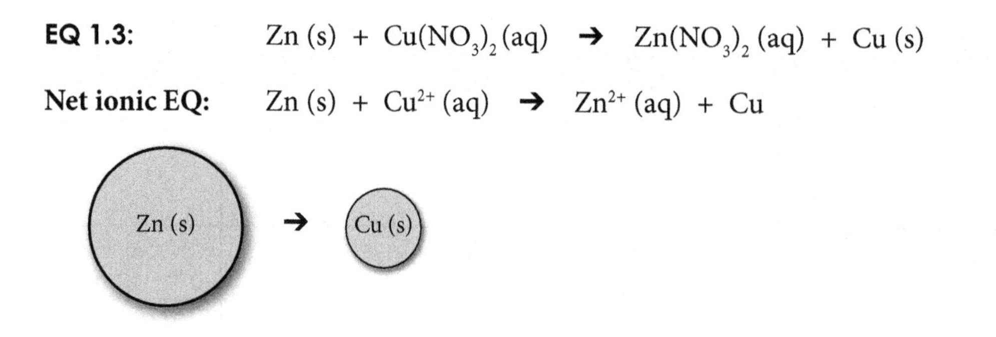

So this reaction occurs because zinc metal is more reactive than is copper. In other words, zinc is more unstable than copper. In drawing the chemical PE's for Zn (s) and Cu (s), we assume that Cu^{2+} and Zn^{2+} aqueous ions have roughly similar chemical PE's.

On the other hand, copper metal does not react to remove Zn^{2+} ion from solution because copper has a much lower chemical PE than zinc. With respect to the chemical potential energies of Cu(s) and Zn^{2+} (aq):

EQ 1.4: $Cu\ (s) + Zn(NO_3)_2\ (aq) \rightarrow$ No Rxn

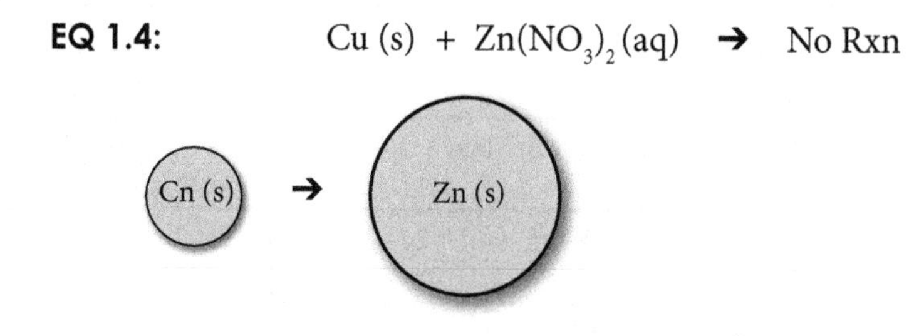

You may still be wondering . . . Why does this reaction in EQ 1.4 <u>not</u> occur? We will discuss this matter in the next section.

Sec 1.2 Law of Conservation of Energy

Chemical potential energy is similar to mechanical potential energy, which everyone probably understands. For example, if a 10-pound lead ball is held 6 feet above the floor, then it has mechanical potential energy. Furthermore, if a 20-pound lead ball is held 6 feet over the floor, then it has twice as much potential energy. If a 10-pound lead ball is held 3 feet above the floor, then it has half of the PE as compared to being held at 6 feet above the floor.

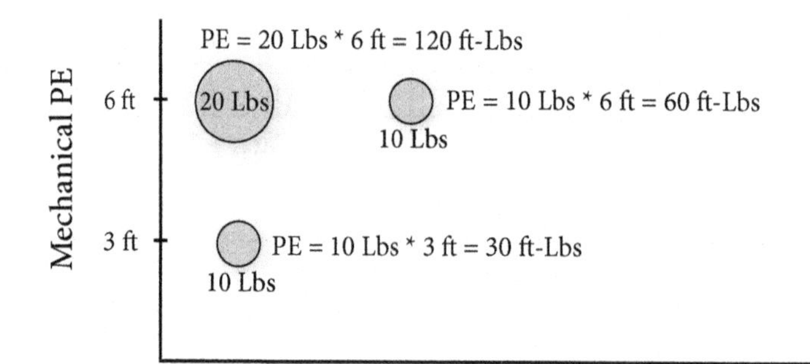

FIGURE 1.1 Mechanical potential energy of lead balls at different heights

As you probably already know, when a lead ball is released, it will have kinetic energy due to its motion (KE = ½ m * v^2), where m is the mass of the ball and v is its velocity (~speed). Furthermore, the *law of conservation of energy* states that the total amount of energy before being dropped equals the energy after it is released. This is also called the *first law of thermodynamics*. Thus, $PE_{before} = KE_{during\ the\ fall} + PE_{height\ of\ the\ falling\ ball}$.

This law also applies to all chemical reactions. The question is where is the energy located with respect to the chemical PE's—Is it on the side of the reactant(s) or product(s)? The answer is that energy is always on the side of the chemical equation with the smaller *chemical PE* (i.e., smaller circles). This is shown for the following reaction:

EQ 1.7: $2\ H_2\ (g) + O_2\ (g) \rightarrow 2\ H_2O\ (l) + \text{energy}$

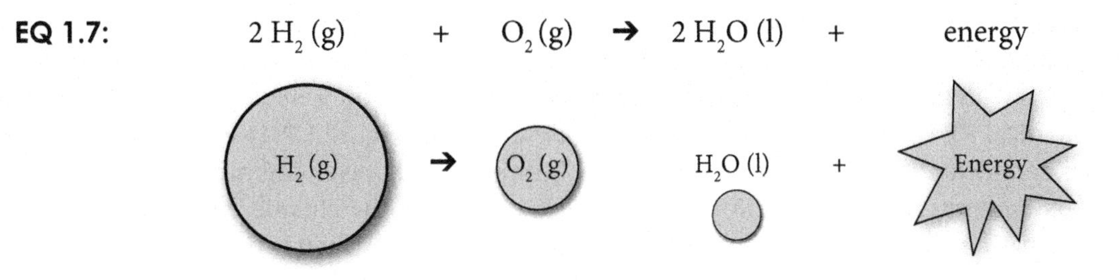

Please note that energy (shown as a "starburst") is always placed on the side with the smaller amount of chemical PE—in this case, the side with water, which is a very stable chemical species.

For this chemical reaction, the chemical PE of the hydrogen gas plus the chemical PE of the oxygen gas is equal to the chemical PE of the water plus the energy that is released, $PE_{H2} + PE_{O2} = PE_{H2O} +$ energy. Please note that the stoichiometry of the balanced chemical equation does not apply to the "energy balance." If the sizes of the three circles and the "starburst" are properly drawn—The areas of hydrogen and oxygen chemical PE's are equal to the areas of water's chemical PE plus the area of the starburst (area of the starburst could be calculated using methods from calculus or geometry). Heat is the most frequent form of energy in chemical reactions. When a balloon is filled with oxygen gas and hydrogen gas (twice the volume of O_2), then the explosion that results is due to the heat energy that is rapidly released. This heat energy comes from the chemical PE's of the two reactants—H_2 and O_2. An *exothermic reaction* occurs when heat is given off by a chemical reaction, and heat energy is always on the product side.

Does the reverse reaction ever occur (see EQ 1.8, below)? The answer is "yes" but only if energy is continually supplied during the reaction. When water (plus electrolytes to catalyze the reaction) is placed in a special type of glassware that has two electrodes connected to a battery, then water is broken down (i.e., decomposed) into hydrogen and oxygen gases. Each gas is collected in a separate tube. In the case of an electrochemical reaction (Chapter 18 in many textbooks), the energy is in the form of electrical energy (Volts). Thus, it is possible to calculate the amount of voltage needed for the decomposition of water to occur. This is an *endothermic reaction* because energy must be continually supplied, and heat energy is always on the reactant side. For EQ 1.8, when the switch on the battery's circuit is turned off, the chemical reaction stops immediately.

EQ 1.8: $2\ H_2O\ (l) + \text{energy} \rightarrow 2\ H_2\ (g) + O_2\ (g)$

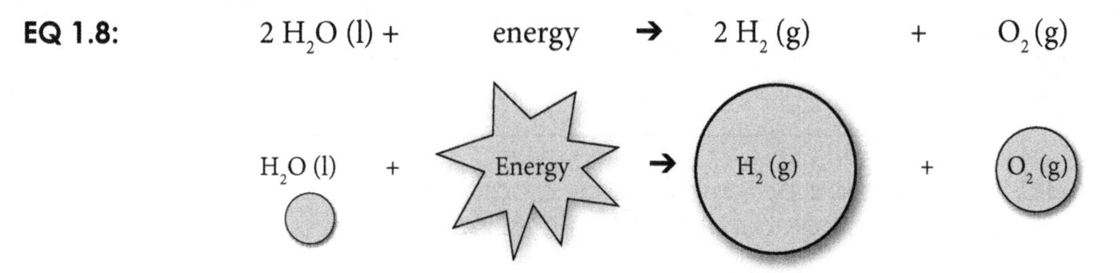

When properly drawn, the areas of the water, and "energy" are equal to the areas of the hydrogen and oxygen gases, PE_{H2O} + Energy = PE_{H2} + PE_{O2}. Remember that the area of a circle represents the chemical potential energy of that particular chemical species. Energy, whether in the form of heat or volts (electrical energy) is always represented by a starburst. Which side always gets the "starburst"? Energy is always on the side that has the more stable chemical species. That is, on the product side for exothermic and reactant side for endothermic reactions.

Sec 1.3 Chemical Potential Energy Diagrams

The chemical potential energy of a particular chemical species is always relative to other chemical species that undergo (or might undergo) similar chemical reactions. Thus, the best way to show these relationships is with a chemical potential energy (y-axis) diagram, which is mathematically analogous to the mechanical PE (Figure 1.1). On the chemical PE diagram (Figure 1.2, below), chemical species with higher chemical PE's are higher on the y-axis than those with lower chemical PE's.

To understand any chemical reaction, it is necessary to know the chemical potential energies of all the participating chemical species. Also, you need to know (or estimate) the energy of activation, which is the energy barrier for a chemical reaction. For example, when hydrogen and oxygen gases are mixed in the proper stoichiometric ratio (i.e., moles H_2 to moles O_2 = 2:1) and ignited by a spark or flame, then liquid water is produced (see Figure 1.3, below). On the other hand, if these gases are mixed with no igniter present, then no reaction occurs. This is true even if we were patient and waited for the reaction to occur over hours, weeks, days, or months. The reason is that the energy of activation, E_{act}, must be supplied to start this chemical reaction.

Referring to Figure 1.3, what exactly is a *progress of reaction diagram*? The y-axis is chemical potential energy, which we have already discussed. The x-axis is "progress of reaction," but what does this really mean? The x-axis is sort of like "time"; however, it is actually a sequence of molecular events: Before rxn ➔ During rxn ➔ After rxn. Thus, one reactant molecule (atom or ion) can undergo a chemical reaction to form the product, while another reactant molecule has not reacted. So you might

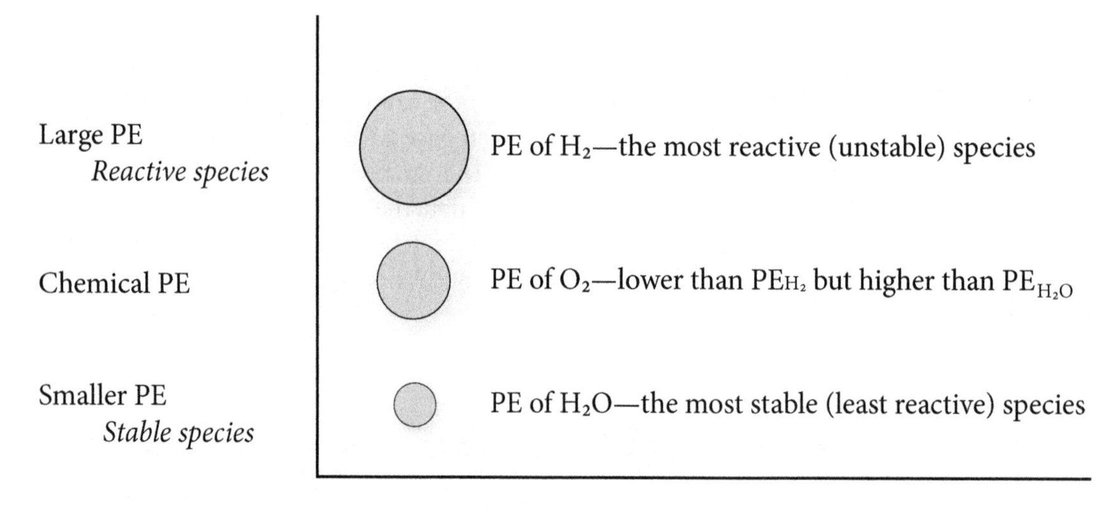

FIGURE 1.2 Chemical potential energy diagram for the chemical species involved in the decomposition of water

say progress of reaction is a timescale for individual molecules, and each molecule has its own time-scale. Don't fret about this distinction—if you want to think of the x-axis as "time of reaction," then you just need to call this as a loosely defined gauge of "time."

How can you identify the *energy of activation* in a "progress of reaction" diagram? First, it is the "energy barrier" to a reaction. To determine its magnitude (size), you extend a horizontal "baseline" at the level of the reactant valley. Next, you draw an arrow from this baseline to the peak of the curve. The energy of activation, E_{act}, is always drawn with an *up arrow*. Why? Energy must always be supplied to overcome this energy barrier, E_{act}. Usually this energy is in the form of collisions between reacting chemical species. This topic will be approached in detail in the Chemical Kinetics unit (Unit 2, and Chapter 12/13/14 for most textbooks).

Sec 1.4 Activation Energy

Very few chemical reactions can occur spontaneously without at least an initial input of energy. For example, a stick of TNT can sit on the floor for years and years and—if undisturbed—there would be no danger of an explosion. Conversely, it explodes immediately when sparked by an ignition source (flame or igniter). The question is then . . . Why doesn't the TNT "self-ignite"? The answer actually involves chemistry—that is, almost every chemical reaction consists of reactant(s) that first must be activated before a chemical reaction can occur. An "energy barrier" is sandwiched between the reactant and product states (see Figure 1.3). This barrier is called the *energy of activation*, $E_{act.}$ A larger E_{act} barrier produces a slower chemical reaction. Nitroglycerin, another explosive, has a very low $E_{act.}$, as compared to TNT's large $E_{act.}$. In fact, nitroglycerin liquid does not need to be ignited by a spark or flame—shake vigorously and it explodes! Overall, reactions that have a smaller energy of activation undergo a more rapid chemical reaction. Although the actual chemical pathways of these two explosives are very different, they both produce explosions when they react. *Explosions* are very rapid chemical reactions that are always exothermic (heat is given off) and always produce a relatively large amount of gaseous products.

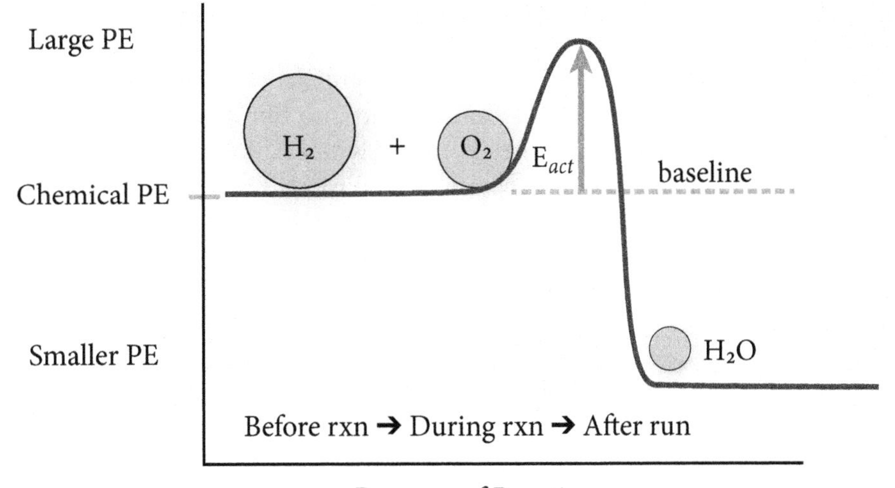

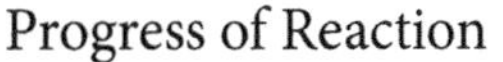

Figure 1.3 Chemical potential energy diagram for the formation of water

When you look at Figure 1.3 (above), you should sense that something is missing—what is it? The answer is that all chemical reactions obey the *law of conservation of energy.* As shown in this figure, there is much more chemical potential energy in the reactants, H_2 and O_2, than there is in the product, H_2O. Thus, this figure is incorrect. To correct it we need to add energy (i.e., a starburst) to the side that has the smaller chemical PE's (see Figure 1.4A). That is, the energy must appear on the product side along with water. So, this reaction is definitely exothermic. For an exothermic reaction, how do you draw the arrow for ΔG_{rxn}*? The arrow begins at the baseline, and extends *downward to the product level* for any exothermic reaction (Figure 1.4A).

Sec 1.5 Chemical Kinetics

Chemical kinetics is the study of factors that affect the rate of a chemical reaction. What factors do you think can affect the rate of a chemical reaction? You probably think first of temperature, pressure (but this is only a factor in some gas phase reactions), concentration of chemical species, and so on. Temperature is definitely a factor, and as a general rule, a 10°C (18°F) increase in the temperature *doubles* the rate of a reaction. You may think that pressure is a factor; however, it is only a factor when there are different total numbers of gaseous moles on each side of the equation: $n_{\text{gas Reactant}} \neq n_{\text{gas Prod}}$. Concentration of chemical species is only a factor for the reactant(s) in a reaction. The concentration of product(s) does not affect the forward rate of reaction. We will deal with this matter in the chemical kinetics unit (Unit 2, and Chapters 12/13/14).

How about an endothermic reaction? For example, water decomposes to form hydrogen and oxygen. For any endothermic reaction, the ΔH_{rxn} arrow always points upward (see Figure 1.4B). Thus, the arrow starts from the baseline and extends *upward to the product level.* Please note that for an endothermic reaction, *E_{act} must be larger than ΔG^o_{rxn}.*

We can use progress of reaction diagrams (Figures 1.4A and 1.4B) to explain how temperature affects the rate of a reaction. At a cooler reaction temperature, the molecules (atoms or ions) are moving slower as compared to their speed at a higher temperature. When a collision between chemical species occurs, chemical bonds may or may not break apart to form the "activated complex," which is found at the peak of the reaction curve. Specifically, if the kinetic energy (KE = ½ m * v²) of the collision is <u>less than</u> the energy barrier, E_{act}, then the reaction will <u>not</u> occur. Conversely, if the KE is greater than the E_{act}, then the reaction <u>may occur</u> but it depends on other factors. A chemical reaction can result when the energetic collision (KE > E_{act}) has the *proper orientation* in space for the molecular species. Say that the "head" of one molecule must bump into the "tail" of another molecule for the reaction to occur. It follows that if the two heads collide with each other, then no reaction will occur. This is true even at elevated temperatures where the collisions possess enough energy (KE > E_{act}). Likewise, if the two tails collide, then no reaction will occur regardless of the temperature.

---

*ΔH_{rxn} and ΔG_{rxn} are different parameters for chemical reactions. However, in this unit we will use them almost interchangeably.

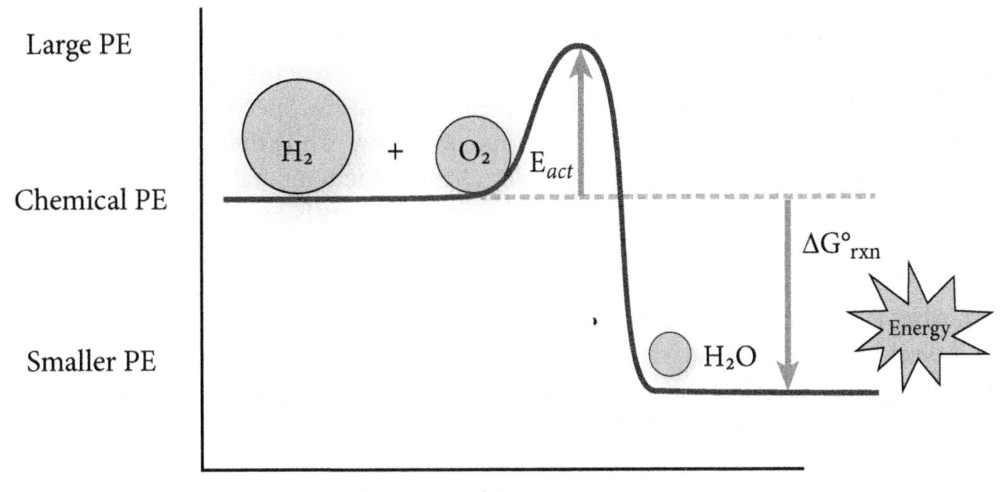

Figure 1.4A Chemical potential energy diagram for the *formation of water*

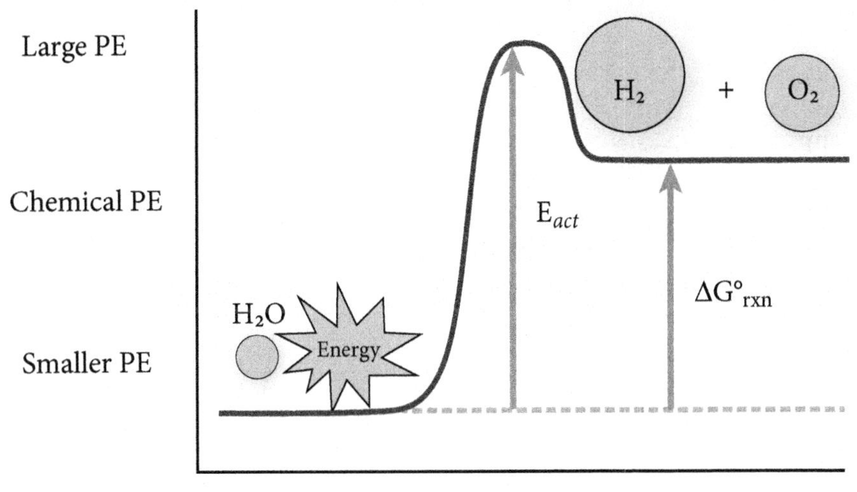

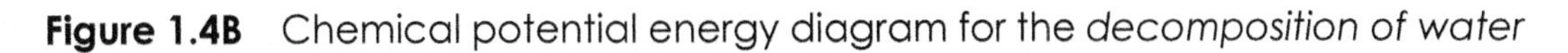

Figure 1.4B Chemical potential energy diagram for the *decomposition of water*

So can we keep it simple for now (the KISS method)? Let's only look at the relationship between the KE of the collision with respect to the E_{act} of the reaction. One thing you should know is that the E_{act} is a constant for a given reaction regardless of temperature. So if a reaction is occurring at a lower temperature, it has the same E_{act} as it does at an elevated temperature. Specifically, in Figure 1.5 (below) the lower temperature is shown by the shorter "blue up arrow," while the elevated temperature is the longer "red up arrow." However, not all molecules at the same temperature are moving at the same speed (velocity). Especially for gas phase chemical reactions, there is a great range of molecular speeds from very slow to very fast. The best estimate of this phenomenon is the average kinetic energy of the molecules. The rule is that the *higher the temperature*, the *greater the average kinetic energy* of the molecules. Also, gas molecules are moving much, much faster than molecules in a liquid. As you will recall gas molecules move in rapid, random directions and speeds, and each molecule is independent of the velocity of nearby molecules. We will discuss the much slower movement of molecules in liquids and solids in the textbook chapter on "intermolecular forces."

Sec 1.6 Chemical Potential Energy of Explosives

Two common explosives are TNT and nitroglycerin. You may wonder . . . Why do these chemicals produce such huge explosions? Chemistry can provide the answer: First, as shown in Figure 1.6, both of these chemical species have very large chemical potential energies. Furthermore, when each one is combusted (i.e., burned), they produce similar sets of gaseous products—carbon dioxide, steam (gaseous H_2O above its boiling point), and nitrogen. As you will recall—for all chemical species, chemical potential energies are relative, but relative to what? If two reactions produce the same products,

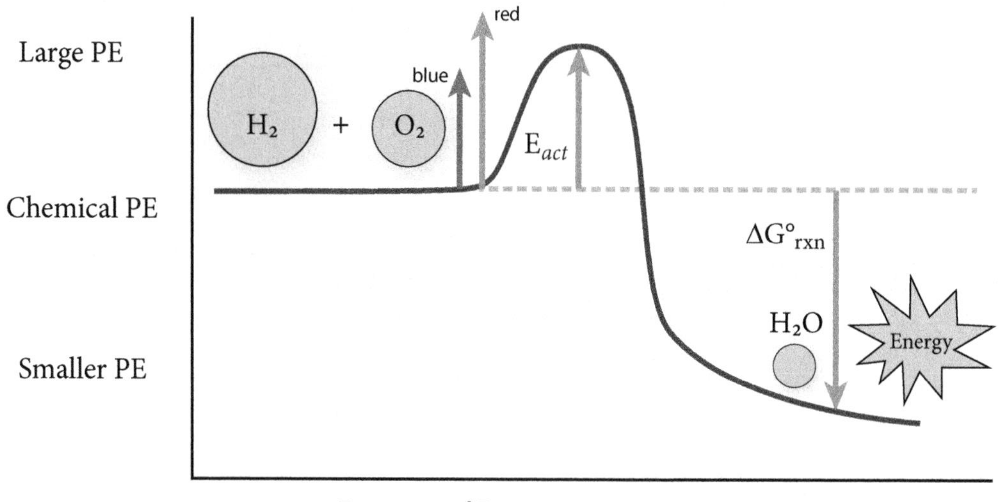

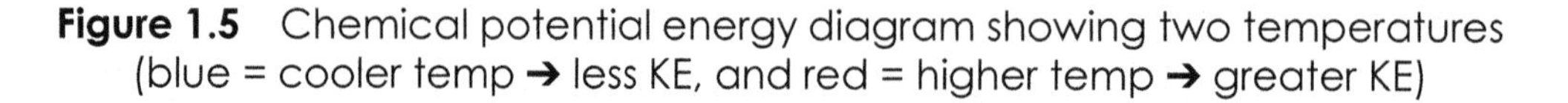
Figure 1.5 Chemical potential energy diagram showing two temperatures (blue = cooler temp → less KE, and red = higher temp → greater KE)

then their progress of reaction diagrams can be directly compared as shown in Figure 1.6. The heat of combustion of TNT (ΔH_{comb} = –3374 kJ/mol) is roughly twice the magnitude of nitroglycerin's (NG) reaction (ΔH_{comb} = –1529 kJ/mol). This fact seems to contradict the fact that NG is a much more dangerous chemical than is TNT. Why? The reason is that the explosive reaction of NG has a much smaller energy of activation (E_{act} = 82 kJ/mol) than the one for the TNT reaction (E_{act} = 195 kJ/mol). Recall that E_{act} is always a positive value, which means energy must be supplied (added) rather than released (lost). It always takes energy to break the chemical bonds in the reactant in order to form the activated species (at the peak of the reaction diagram). Another difference between these two combustion reactions is that NG decomposes by itself without the need for oxygen gas. As we will discuss in the thermodynamics unit (Unit 7 or Chapter 17/18 in your textbook), when an exothermic reaction produces gases, then it tends to happen "spontaneously" (i.e., without the need for energy to be continually supplied). All explosions are *spontaneous chemical reactions* that produce heat energy (i.e., exothermic) and large amounts of gaseous products. If the change in "chemical potential energy, ΔG_{rxn}" is negative, then the reaction is spontaneous.

Conversely, if the change, ΔG_{rxn}, is positive, then the reaction is *nonspontaneous*—energy must be continually supplied to keep the reaction going. For example, when water is decomposed to form hydrogen and oxygen gases (Figure 1.4B)—it takes a constant input of energy from a battery to make this nonspontaneous reaction occur. The moment the battery is disconnected, the tube for each gas, H_2 and O_2, stops bubbling. Another example is from biology—photosynthesis is a nonspontaneous reaction. Continuous input of sunlight is needed for this reaction to occur.

EQ 1.9: $$6\,CO_2\,(g) + 6\,H_2O\,(l) \rightarrow C_6H_{12}O_6\,(aq) + 6\,O_2\,(g)$$

Which side of the equation should contain energy (sunlight)? If you are thinking that it is the left side, then that is correct! Both carbon dioxide and water are very stable molecules—as compared to glucose (fuel for the cell) and oxygen gas. Thus, the correct chemical equation for this reaction is as follows:

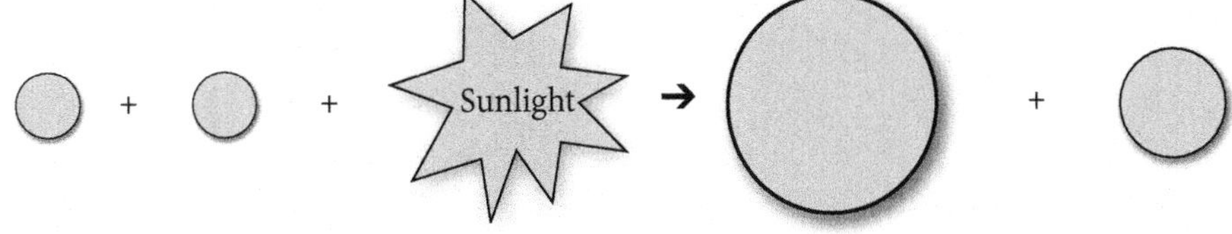

EQ 1.9: $$6\,CO_2\,(g) + 6H_2O(l) + \text{Energy} \rightarrow C_6H_{12}O_6\,(aq) + 6\,O_2\,(g)$$

Furthermore, if the circles and the starburst are properly drawn, then the total area on the left side is equal to the total area on the right side. Also, qualitatively we know that ΔG_{rxn} is positive for this reaction ($\Delta G_{rxn} > 0$)—continuous energy input is required for the reaction to occur. You will learn how to calculate the quantitative values for these reactions when we study thermodynamics (Unit 7 and textbook chapter 17/18).

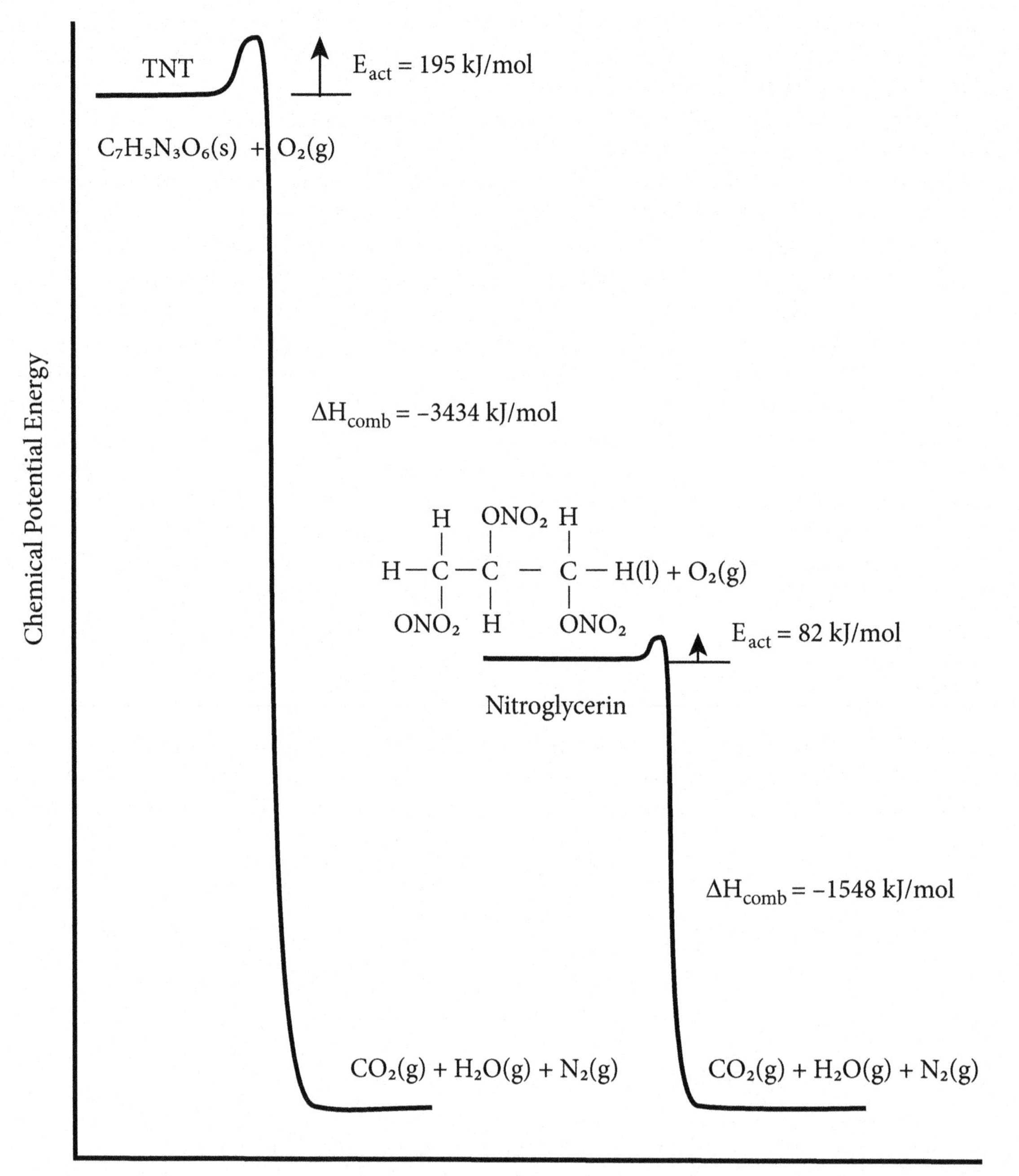

Figure 1.6 Combustion reactions of TNT and nitroglycerin (NG).
Source of information: Walters, 2002[2]

Sec 1.7 Oxidation and Combustion

Combustion reactions, as discussed above, are all oxidation reactions. Why? What is being oxidized? First, the best way to understand the difference between oxidation and reduction is to determine the oxidation number of each chemical species (atom or ion) in a reaction. So, *oxidation* is the process in which the oxidation number of the atom/ion increases in a chemical reaction. For example, when graphite (e.g., pencil lead) reacts with oxygen gas, the carbon undergoes complete oxidation, which produces carbon dioxide gas.

EQ 1.10: $C(s) + O_2(g) \rightarrow CO_2(g)$

As you probably learned in first semester chemistry, *oxidation number* (also called oxidation state) the charge on one atom or ion. The oxidation number of an element in its natural state is always zero, so C(s) is zero. For carbon dioxide, CO_2, the oxide is –2; and to calculate the oxidation number of carbon is just an algebra problem: x + 2 *(–2) = 0, and x = 4+ for carbon. Thus, this process of change in oxidation number of carbon from zero to 4+ is an oxidation. However, you probably recall from first semester chemistry that every oxidation must be accompanied by a reduction. In the case of EQ 1.10, it is oxygen that gets reduced: from zero to –2. Since both oxidation and reduction are occurring together, why is this reaction called an *oxidation*? The answer is that chemists and biologists tend to focus on carbon, and the change in oxidation number of carbon. Overall, combustion is the rapid reaction of any substance with oxygen gas accompanied by the oxidation of carbon (increase in oxidation number).

Sec 1.8 Electronegativity Values for Nonmetal Elements

First, recall that carbon is more electronegative than hydrogen; thus, carbon in methane will have a negative oxidation number and hydrogen a positive number. The electronegativity values of selected nonmetallic elements are shown in Table 1.2 on the next page.

Sec 1.9 Fuels and Chemical Potential Energy

Fossil fuels (such as oil, gas, and coal) are all deposits in the earth. All fuels are heavily reduced. What does this mean? In *methane* (the principal ingredient in natural gas), carbon is fully reduced. It can be formed when carbon (graphite) reacts with hydrogen gas:

EQ 1.11: $C(s) + 2\,H_2(g) \rightarrow CH_4(g)$

What are the possible oxidation numbers of hydrogen? Hydrogen can be neutral (hydrogen's natural state is H_2 (g)), positive (e.g., HF fluorine is more electronegative than H), or negative (e.g., NaH where sodium has a lower electronegativity). So, in EQ 1.11 the oxidation number of hydrogen goes from zero to plus one (i.e., oxidation). We can calculate the oxidation number of carbon as –4 as shown by the following: x + 4 *(+1) = 0, x = –4. Thus, in the formation of methane, carbon goes from zero to –4, which is reduction. In all fuels, carbon is fully or partially reduced. Another example is CH_3OH

Table 1.2 Shortened Periodic Table designed to show the electronegative values of selected nonmetal elements, and the application of the octet rule to nonmetal atoms in different chemical periods.

Period 1: Electronegativity	**H** **2:1**						**He** **—**
Period 2: Electronegativity		**B** **2.0**	**C** **2.5**	**N** **3.0**	**O** **3.5**	**F** **4.0**	**Ne** **—**
Stability	*1 bond*	*3 bonds*	*4 bonds*	*3 bonds*	*2 bonds*	*1 bond*	*None (monoatomic)*
Stable structures	H– H—F: O=C(H)H	—B— :F—B—F: (with :F:)	—C— CH_4 (H—C—H) O=C(H)H H—C≡N:	—N— H—N—H (with H) :N≡N:	—O: H—O—H O=C(H)H	—F: H—F: :F—F:	:Ne:

(methanol or methyl alcohol), which is the fuel used in some drag-racing vehicles. What is the oxidation number for carbon in methanol? Oxide is –2 and hydrogen is +1, so carbon is –2 (x +3*(+1) + (–2); x = –2 (or 2–). Why is methanol a good fuel? It can undergo complete oxidation as follows:

EQ 1.12: $CH_3OH\ (l) + 3/2\ O_2\ (g) \rightarrow CO_2\ (g) + 2\ H_2O\ (l)$

How do you predict whether the reaction in EQ 1.12 is exo- or endo-thermic? You should be able to answer this question in a qualitative manner—but how? Methanol has a very high chemical potential energy because it is relatively unstable (i.e., reactive). Oxygen gas has an "intermediate" chemical PE as shown when a fire burns, it is oxygen combining with the "fuel" not nitrogen. Carbon dioxide and water are both very stable molecules—thus, they have very small chemical PE's. Which side does heat energy go on? Remember that energy always goes on the side with the stable molecules. So we can draw EQ 1.12 as follows:

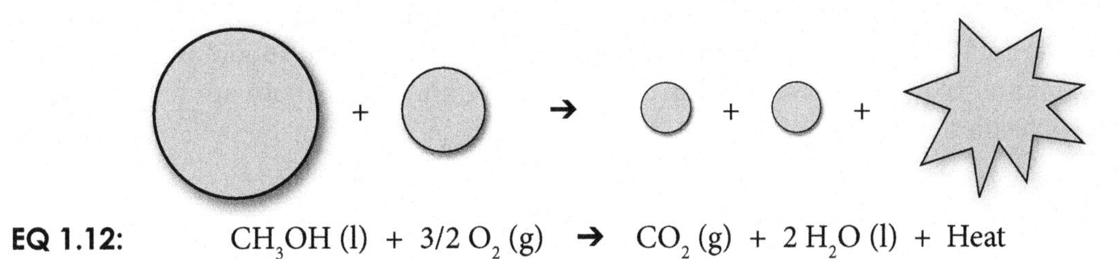

EQ 1.12: $CH_3OH\ (l) + 3/2\ O_2\ (g) \rightarrow CO_2\ (g) + 2\ H_2O\ (l) + \text{Heat}$

Using the Law of Conservation of Energy, if the circles (chem PE's) and starburst (heat) are properly drawn, the total area on the left side (reactants) will equal the total area on the right side (products). So, the combustion of methanol is exothermic. In fact, all combustion reactions of fuels are exothermic.

Is it possible to determine the relative fuel value of different fuels? Yes, for fuels containing carbon in a reduced state, we can use oxidation numbers to determine their relative chemical potential energies (fuel values). Methane, CH_4 (g), is a better fuel than is methanol, CH_3OH (l). This is due to the oxidation number of carbon being 4– in methane, while it is 2– in methanol. However, there is an easier way to determine chemical potential energies of fuels. The more hydrogen a carbon atom has, the greater its chemical PE and hence its fuel value. So, looking at the following sequence from largest chemical PE to smallest (where double-bond is shown as C=O):

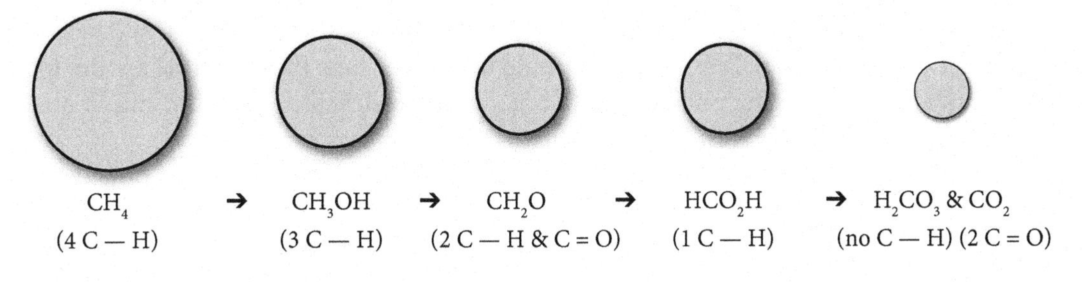

CH_4 (4 C — H) → CH_3OH (3 C — H) → CH_2O (2 C — H & C = O) → HCO_2H (1 C — H) → H_2CO_3 & CO_2 (no C — H) (2 C = O)

Do you see another factor that could be used to determine this sequence? It is that fuel value (chemical PE) decreases with the addition of oxygen. Thus, removal of hydrogen and/or addition of oxygen to carbon both decrease the chemical PE of carbon-containing compounds.

What fuel is used in living cells? Glucose, $C_6H_{12}O_6$, is the fuel used by both animal and plant cells. Unfortunately, this formula obscures the number of H's and O's on each carbon. The carbons in glucose are bonded together by single bonds and the bonding on each carbon is H—C—O—H. Note that each carbon carries only one hydrogen (C—H), thus glucose is classified as a fuel for plant and animal cells. When a cell needs energy from glucose (stored as ATP), glycolysis strips hydrogen off of carbon (while giving H's to NAD^+). That is, the C—H bond provides the energy in glucose that is transferred to ATP. Overall, which fuels in the above sequence are better fuels than glucose? The better fuels are . . . methane (4H), methanol (3H), and formaldehyde (2H). In your biology classes, you will probably learn that there are bacteria that metabolize methane as well as those that metabolize methanol. Which molecule has a greater chemical potential energy? As shown in Figure 1.7, both methane and methanol undergo combustion to produce carbon dioxide and water. Given their chemical PE's, which combustion is more exothermic? Methane's reaction has a larger ΔG_{rxn} because it has greater chemical PE and both reactions produce the same products. Furthermore, formaldehyde (2H) is an intermediate formed during the metabolism of methanol. These bacteria have adapted to using a fuel that has greater fuel value than organisms that use glucose. Overall, you should remember that the complete oxidation of carbon always yields carbon dioxide and water (both are fully oxidized and very stable molecules).

Sec 1.10 Chemical Bonding of the Common Nonmetal Elements

How do chemists know whether or not an element is stable within a particular molecule? The answer to this question is based on the electron configurations of valence electrons, which you studied last semester. The number of chemical bonds (i.e., covalent bonds) needed to make stable atoms varies in a periodic manner for the common nonmetal elements as shown in Table 1.2 (previous section). As you learned last semester, the noble gases are stable as monoatomic species; that is, with no bonds to any other atom. The halogens are stable with one covalent bond (e.g., H—Cl), oxygen with two bonds (e.g., H—O—H, or C=0), nitrogen with three bonds (e.g., NH_3, N_2), carbon with four bonds (e.g., CH_4, $H_2C{=}CH_2$, HCCH has a triple bond between C's), and boron with three bonds (e.g., BF_3). So if the number of covalent bonds differs from this pattern, then the molecules are usually highly unstable (e.g., CH_3· or H_2N:). The reason for the stability pattern of bonds is the *octet rule*—something you learned last semester. An atom or ion is stable when it has eight valence electrons. Hydrogen obeys the duet rule, and boron can be stable with six, plus nonmetal atoms below the second period can expand their octet (e.g., PCl_5 SF_6 or XeF_4). So you should remember how the octet rule applies to the nonmetals throughout this semester and beyond to organic chemistry, biochemistry, and so on.

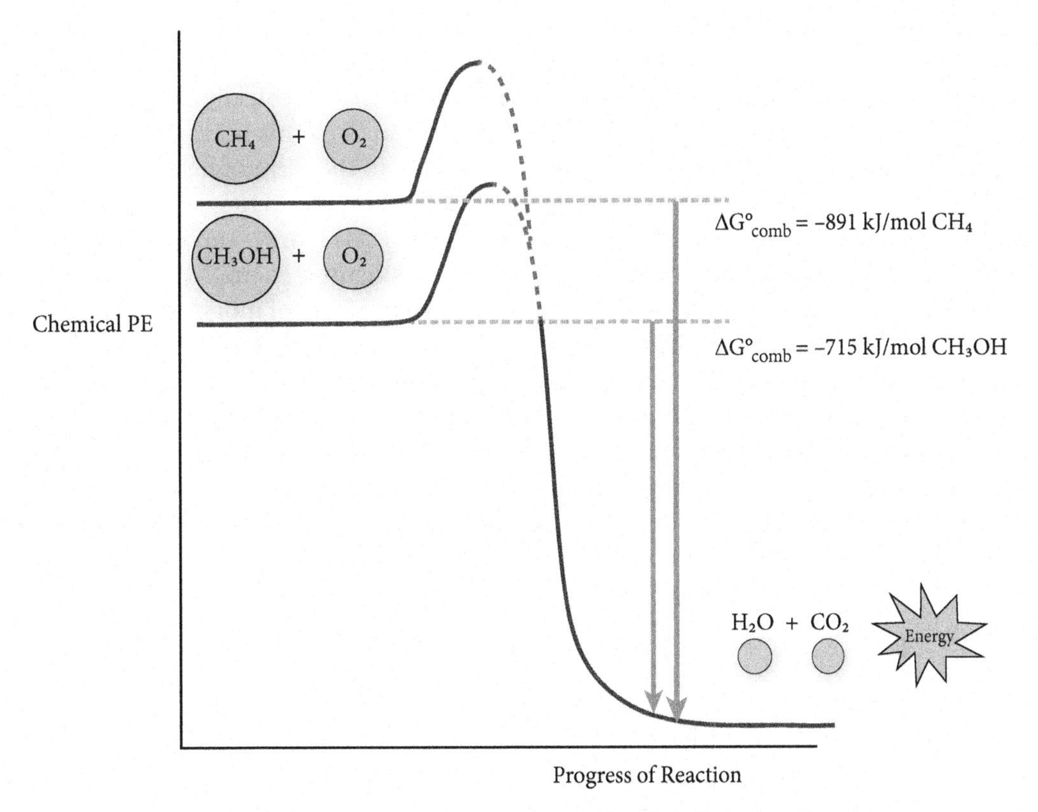

Figure 1.7 Chemical potential energy diagram for the *combustion of methane and methanol*

Sec 1.11 Does Breaking Bonds Produce Energy?

There are always two phases in a chemical reaction: first, chemical bonds of the reactant(s) are broken, then new bonds are formed to make the product(s). So it is a *misconception* that breaking bonds releases energy. *Breaking bonds always costs energy.* This energy must come from fast moving molecules (atoms or ions) that collide with each other. These collisions must be highly energetic and they must have the proper orientation in order to break bond(s) and to form the activated complex at the peak of the reaction curve. Next, when the activated complex (peak) forms new bonds, then the product(s) are made. Bond making always releases energy. Overall, if the cost of breaking bonds is greater than the energy released when bonds are made, then the reaction will be endothermic. Conversely, if the energy released is greater than the cost of breaking bonds, then the reaction is exothermic. So this can be thought of as:

ΔH_{rxn} (net energy) = (Energy cost of breaking bonds) – (Energy released by new bonds)
ΔH_{rxn} (net energy) = (negative) kJ/mol ➔ Exothermic Reaction ($\Delta H_{rxn} < 0$)
ΔH_{rxn} (net energy) = (positive) kJ/mol ➔ Endothermic Reaction ($\Delta H_{rxn} > 0$)

How can you remember these concepts, which are counter-intuitive? Get two *magnets* that are stuck together. Try pulling them apart (if you pump iron, then just pretend) and note that it takes energy to separate them. Likewise, it takes energy to separate atoms that are chemically bonded together. Now hold the two magnets very close together. Feel the *force of attraction*. This is analogous to new chemical bonds being made, which always releases energy—usually in the form of heat.

Have you ever gone snow skiing? If "yes," imagine when grandpa used to go skiing. He had to walk up the snow-covered hill (uses KE ➔ increase PE) carrying his skis (sorry—no ski lifts in them days). This is work but he is building up potential energy—analogous to a chemical reaction where molecules (ions or atoms) go uphill to the peak. It takes energy in the form of moving molecules that must collide—an energetic collision with the proper orientation of molecules. Next, as shown in Figure 1.8 (below), grandpa smiles, puts on his skis, and down the snow-covered mountain he goes (wheee!!!). The downhill skiing part (uses PE ➔ increase KE) is analogous to the formation of bonds, which releases energy. First, imagine that grandpa trudges up a small hill and then he skis down a very steep slope ☺. What type of chemical reaction is analogous to this change in mechanical potential energy? Did you think "exothermic"? ☺ Next, grandpa trudges up a very steep mountain, puts on his skis, and then skis down a very small hill—bummer ☹. Which type of chemical reaction? Endothermic.

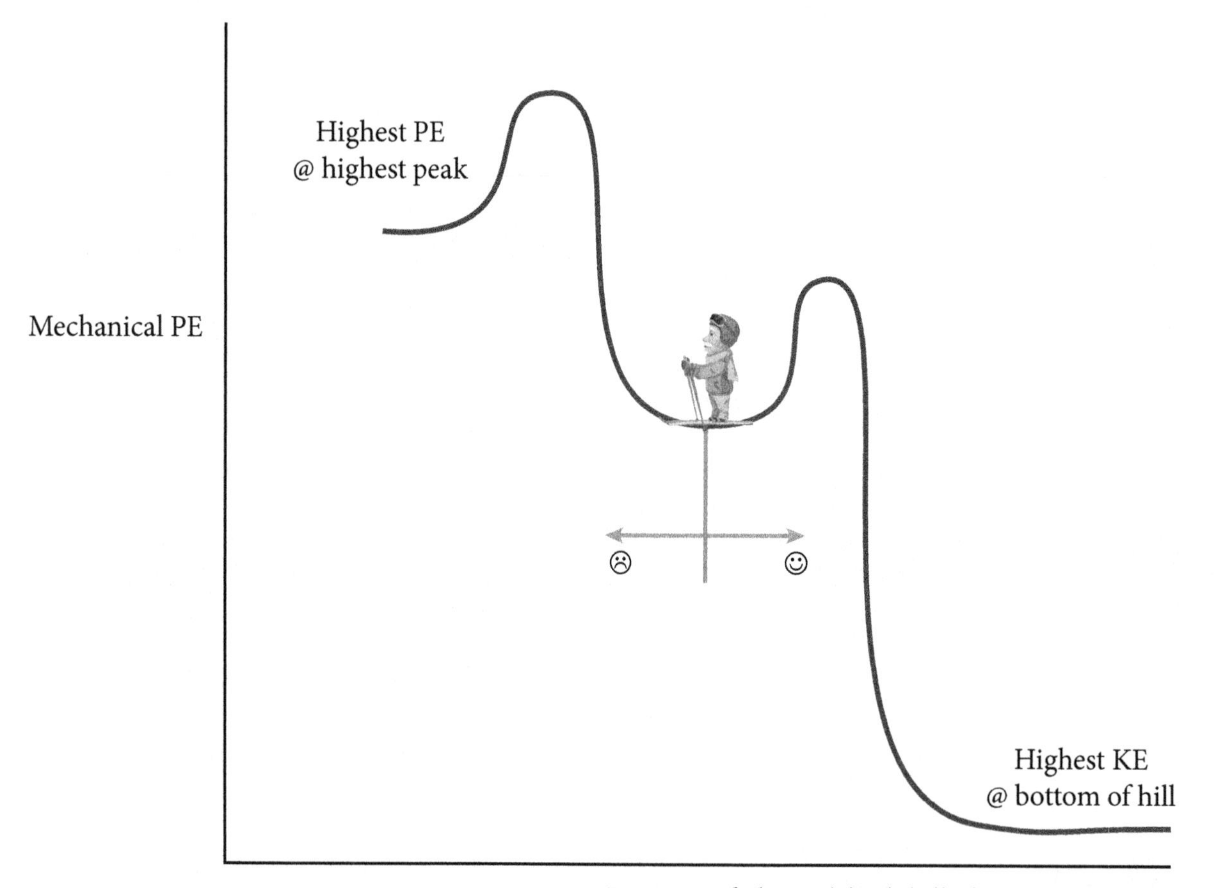

Figure 1.8 Mechanical potential energy diagram:
Two mountain ski slopes—Which hill is exo-? . . . and the other is endo-

Sec 1.12 Summary

Understanding chemical potential energy is an essential skill in chemistry courses—from second semester general chemistry to organic chemistry to biochemistry (and finally analytical and physical chemistry plus molecular biology). The skill of balancing a chemical equation does not help one determine if a chemical reaction should occur. Rather, an understanding of chemical potential energy is needed to determine if a reaction should occur. If it occurs, then whether it is exothermic or endothermic depends on the chemical potential energies of the reactants and products. Chemical PE's are described on a *"per mole" basis* (kJ for one mole), which means that the balanced chemical equation is not an appropriate way to gauge the "energetics" of chemical species (i.e., molecules/atoms/ions). Furthermore, in Unit 1 we are not concerned with calculating numerical results for the overall Gibbs free energy of a reaction, ΔG_{rxn}. This quantitative topic will be addressed in the unit on thermodynamics (Unit 7) and the textbook chapter (17 or 18).

Suppose that two chemical species undergo similar reactions and they produce the same products. If the reaction of one species yields a large energy change, ΔG_{rxn}, does that mean it will be the faster reaction? No! The reaction with the smaller energy of activation, E_{act}, will undergo the faster reaction. In fact when comparing two explosive chemicals—TNT and nitroglycerin (NG)—TNT has a much larger ΔH_{rxn} than does NG. But NG has a much lower E_{act} than does the TNT combustion. All combustion reactions are oxidation reactions—because we focus on carbon, and carbon is oxidized in all combustions [e.g., C^{o} (s) ➔ C^{4+} in CO_2 (g)]. It is important to be able to determine the relative stability of a molecule (ions or atoms). Why? For example, you should know that all exothermic reactions proceed from unstable (reactive) to stable molecules and energy (starburst) is always shown on the product side. This sequence is reversed for endothermic reactions.

It is important to use the electronegativity values (Table 1.2) to determine the oxidation number in a chemical species—if we focus on fuels: fuels are molecules that have very high chemical potential energy because they are fully or partially reduced. For example, methane (CH_4, where C is -4) is a better fuel than is methanol (CH_3OH, where C is -2), and methanol is a better fuel than is glucose (H—C—O—H, the C is -1 with respect to its lone H). However, remember that there is an easier way to determine fuel value—the more H's bonded directly to C, the greater its fuel value (i.e., chemical PE). When you are asked to determine the stability of a molecule, remember how the octet rule applies to the number of chemical bonds in nonmetal atoms that are needed to produce stable molecules (Table 1.2). Overall, an understanding of chemical reactions requires knowing that it costs energy to break chemical bonds, while forming bonds releases heat energy. If the "energy cost" (bond breakage) is greater than energy released when the new bonds are formed, then the reaction is *endothermic*. Likewise, when the *energy released* when new bonds are formed is greater than the cost, the reaction is *exothermic*. Remember the analogy with grandpa's ski stories—he had to walk up snow-covered hills (energy "cost") before he could ski down the mountain!

Sec 1.13 Technical References

Taken from free-access journals, and intended mostly for your chemistry professor's use.

[1] Kleinman, R. W., Griffin, H. C., & Kerner, N. K. Images in chemistry. *Journal of Chemical Education*, **1987**, *64* (9), 766–770.

[2] Walters, R. N. Molar group contributions to the heat of combustion. *Fire and Materials*, **2002**, *26:* 131–145.

Sec 1.14 Questions to Ponder

Q1A. Using Table 1.1, which one of the following equations represents a chemical reaction that does not actually occur?

X. Mg (s) + H_2O (g) → ??
Y. Ni (s) + $Fe(NO_3)_2$ (aq) → ??
Z. Fe (s) + $Ni(NO_3)_2$ (aq) → ??

1B. Explain your answer in terms of chemical potential energies.

1C. Write the balanced net ionic equation (you may have to review this first semester topic) for the two equations (X/Y/Z) where the reaction actually occurs.

Q2A. Which allotrope of oxygen should have the higher chemical potential energy? Briefly explain your answer.

X. O_2 (g)
Y. O_3 (g)

2B. Write the balanced chemical equation for the formation of ozone from O_2 (g).

2C. Use this equation (2B) to show the chemical potential energies as "circles" and energy as "star-burst." Is this reaction exo- or endo-thermic?

Q3A. Look at these two chemical reactions. For each one, is it an endothermic or exothermic reaction?

X. N_2 (g) + 3 H_2 (g) → 2 NH_3 (g) $\Delta H_{rxn} = -45.90$ kJ/mol
Y. N_2 (g) + 3 I_2 (s) → 2 NI_3 (s) $\Delta H_{rxn} = +290$ kJ/mol

3B. As you may recall from first semester chemistry, when you switch reactants and products, you change the sign on the enthalpy of reaction, ΔH_{rxn}. Based on this limited amount of information, which *reverse reaction* is most likely an explosive reaction? Write the chemical equation, including ΔH_{rxn}, for this particular reverse reaction.

3C. Draw the chemical potential energies (y-axis) and progress of reaction (x-axis) for the reverse reaction in 3B. Hint: this reaction has a very low E_{act}. What is the significance of this fact?

3*. (Real World Question—just for fun and not points) Which two of the chemical species in the above equations (X and Y) have household uses? Hint: these are relatively mild chemicals (relatively stable).

Q4A. When adenosine triphosphate (ATP) undergoes hydrolysis (reaction with water) it produces adenosine diphosphate (ADP) and phosphate ion (P_i).

ATP (aq) + H_2O (l) → ADP (aq) + P_i (aq)

The Gibbs free energy for this reaction is: $\Delta G_{rxn} = -30.5$ kJ/mol. Ignoring phosphate ion, which molecule is more stable—ATP or ADP?

4B. In this reaction, does the energy involved come from "bonds made" or "bonds broken"?

4C. Draw the chemical potential energy (y-axis) versus progress of reaction (x-axis) for this reaction. Use circles to represent chemical PE's and starburst for energy.

Q5A. With respect to the octet rule and number of bonds per nonmetal atom: Which one of the following molecules is <u>unstable</u> (actually it does not exist with the bonding shown)? Please note that H_2X (where X is C or N atoms) means the two hydrogen atoms are directly bonded to X (C or N).

X. $H_2C{=}O$ (g)

Y. $H_2C{=}CH_2$ (g)

Z. $H_2N{=}NH_2$ (l)

5B. If you answered correctly on 5A, then you may find it interesting that this molecule does <u>not</u> exist as written above. That is the double bond must be replaced with a single bond—the resulting molecule is actually a *rocket fuel*! Also, it has been found in cigarette smoke in small amounts. Write the reaction of this molecule when it forms a more stable molecule. Hint: This "more stable" molecule is a common household chemical and you have probably heard of it before taking chemistry.

5C. Draw the chemical potential energy (y-axis) versus progress of reaction (x-axis) for the reaction in 5B. Use circles to represent chemical PE's and starburst for energy.

UNIT 2

Visualizing Chemical Reaction Rates

WHY STUDY THIS UNIT?

The goal of this unit is to help you develop a qualitative understanding of the factors that affect chemical kinetics, which is the study of reaction rates. If you can learn to visualize how these factors affect rate, then it will be much easier to learn how to do the appropriate mathematical calculations. The corresponding chapter in textbooks is entitled "Chemical Kinetics," and it is usually Chapters 12, 13, or 14 in most textbooks.

Sec 2.1 Introduction

As briefly discussed in Unit 1 (Sec 1.5), *chemical kinetics* is the study of factors that affect the rate of a chemical reaction. In this unit we will explore these factors in depth. A common chemical reaction that you may be familiar with is the decomposition of hydrogen peroxide to form water and oxygen gas. *Hydrogen peroxide* is used as a medicine (apply to cleanse a cut or wound) or as the first step in coloring hair (H_2O_2 bleaches the hair to "bleach blonde," then apply desired hair color solution). The chemical equation for this decomposition reaction is shown below:[1–4]

EQ 2.1: $$H_2O_2\,(l) \rightarrow H_2O\,(l) + O_2\,(g)$$

How do we know the rate at which this reaction occurs? To monitor this rate, you can plot a graph of the rate of gas formation, or the rate of disappearance of hydrogen peroxide.

When a chemical reaction occurs, the *initial rate of reaction* is determined using data collected at the very start of the reaction. As shown in the graph for the decomposition of hydrogen peroxide (see Figure 2.1), the rate is calculated from the slope of the time (x-axis) versus concentration (y-axis) graph. Does this rate stay the same over the entire reaction? No, obviously as shown by the change in slope from steep to shallow, the rate decreases with time of reaction. The reason for this decrease is that the amount of reactant, in this case H_2O_2, is being consumed and hence less and less of it is available to react (i.e., decompose). You probably recall how to determine the slope of a line from math classes: slope = m = $(\Delta y/\Delta x) = (y_2 - y_1)/(x_2 - x_1)$. First, you should find a point on the curve where you want to know its slope. Next, draw a straight line that only goes through that point (i.e., tangent to the line). On this line, pick any value of x for point 1 (x_1, y_1) and point 2 (x_2, y_2) where point 2 is always to the right of point 1. For initial rate, calculate the slope that goes through the origin (x = t = 0). As shown in Figure 2.1, the initial rate of reaction is $R = -3.85 * 10^{-4}$ M s^{-1}. The negative sign indicates that the value of y is decreasing with x (i.e., negative slope). As described in the next paragraph, chemists always use the initial rate of reaction to compare how different factors affect the rate of reaction.

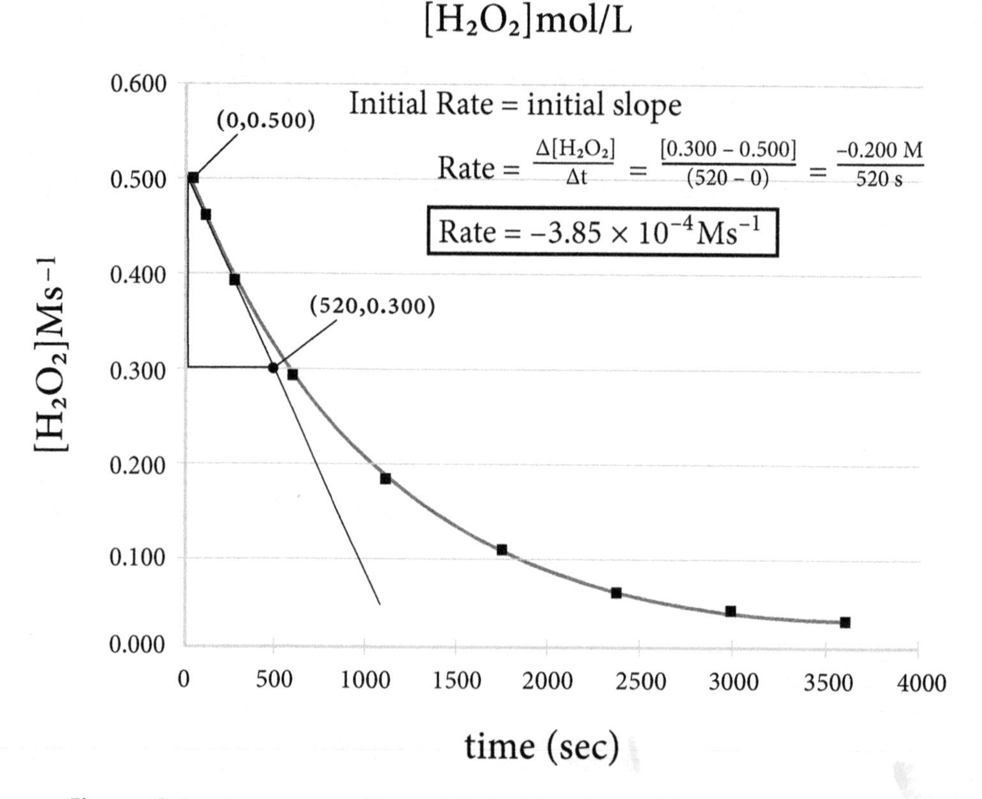

Figure 2.1 Decomposition of H_2O_2 Kinetics: Initial Rate of Reaction

As described above, concentration of chemical species is definitely a factor in determining reaction rate (see Figure 2.2). The concentration of product does not affect initial rate because its initial concentration is zero. Overall concentration of product(s) does not affect the *forward rate* of reaction; however, when we cover the unit on chemical equilibrium, product concentration affects the rate of the reverse reaction.

What other factors affect reaction rate? Obvious answers are temperature, pressure (for gases only), presence of a catalyst, and so on (see Table 2.1). Temperature is definitely a factor, and as a general rule, a 10°C (18°F) increase in the temperature *doubles* the rate of a reaction. You may think that pressure is a factor; however, it is only a factor when there are different total numbers of gas moles on each side of the equation: $n_{gas\ Reactant} \neq n_{gas\ Prod}$.

Sec 2.2 Order of a Reaction

Looking at a graph of concentration (y-axis) versus time of reaction (x-axis) is probably the easiest way to understand "order of a reaction." If the plot is a straight line with a negative slope, then the reaction is zero-order (see Figure 2.3). Why "zero-order"? What does it mean? If a particular reactant gives a zero-order plot, then the rate of reaction does not depend on the concentration of the reactant, [R]. That is, the rate is the same for the entire time of reaction. Recall that the slope of [R] versus time is equal to the rate of reaction. Thus, a straight line means the rate is the same regardless of the concentration of reactant.

Table 2.1 Factors that affect reaction rates

Factor	Increases Rate	Decreases Rate	Comments
Temperature	Raise by 10°C (18°F) ~ doubles rate	Lower by 10°C (18°F) ~ cuts rate in ½	This is a general rule that has exceptions
Pressure	Increases—if fewer gaseous product than gas reactants	Decreases—if more gaseous product than gas reactants	Balanced chemical equation
Concentration of Reactant	Doubling [Reactant], doubles rxn rate		First-order rxn
"	Doubling [Reactant], quadruples rxn rate		Second-order rxn
"	Doubling [Reactant] has no affect rxn rate		Zero-order rxn
Catalyst	Lower E_{act} ~ Faster rxn	N/A	Enzymes: biological catalysts, which are very specific
Inhibitor (negative catalyst)	N/A	Raises E_{act} ~ Slower rate	
* ΔH_{rxn} and ΔG_{rxn} are different parameters for chemical reactions (see Unit 7). However, in this unit we will use them as if they are the same parameter.			

For both first- and second-order reactions, the graph of concentration of reactant [R] versus time is a curve with a negative slope. When both reactions produce the same type curve—how can you determine the order of reaction? The answer is to use various mathematical transformations on the y-axis to search for a plot that produces a straight line. Specifically, when data is plotted as natural log of concentration, ln [R], versus time (see Figure 2.2)—a straight line with a negative slope results for a *first-order reaction.* Meanwhile, the plot of ln[R] versus time for a second-order reaction produces a curved line. Conversely, when the inverse of concentration, 1/[R], versus time produces a straight line with a positive slope (see Figure 2.4), the *reaction is second-order.* The plot of 1/[R] versus time for a first-order reaction is a curve with a set of positive slopes.

Your textbook probably carefully describes the relationship between rate of reaction, R, and concentration of reactant, [R], in great detail. This relationship always depends on the order of reaction as follows:

First-order rxn: $R \propto [R]^1$, → doubling the concentration of R, <u>doubles</u> the rate

$R_1 \propto [1]^1$, $R_2 \propto [2]^1$, and $R_2 = 2\,R_1$

Second-order rxn: $R \propto [R]^2$, → doubling the concentration of R, <u>quadruples</u> the rate

$R_1 \propto [1]^1$, $R_2 \propto [2]^2$, and $R_2 = 4\,R_1$

Zero-order rxn: $R \propto [R]^0$, → doubling the concentration of R, results in the <u>same rate</u>

$R_1 \propto [1]^1$, $R_2 \propto [2]^0$, and $R_2 = R_1$

Decomposition of $H_2O_2(l)$ **What is the rate constant, k, for this reaction?**
What is the half life for this reaction?

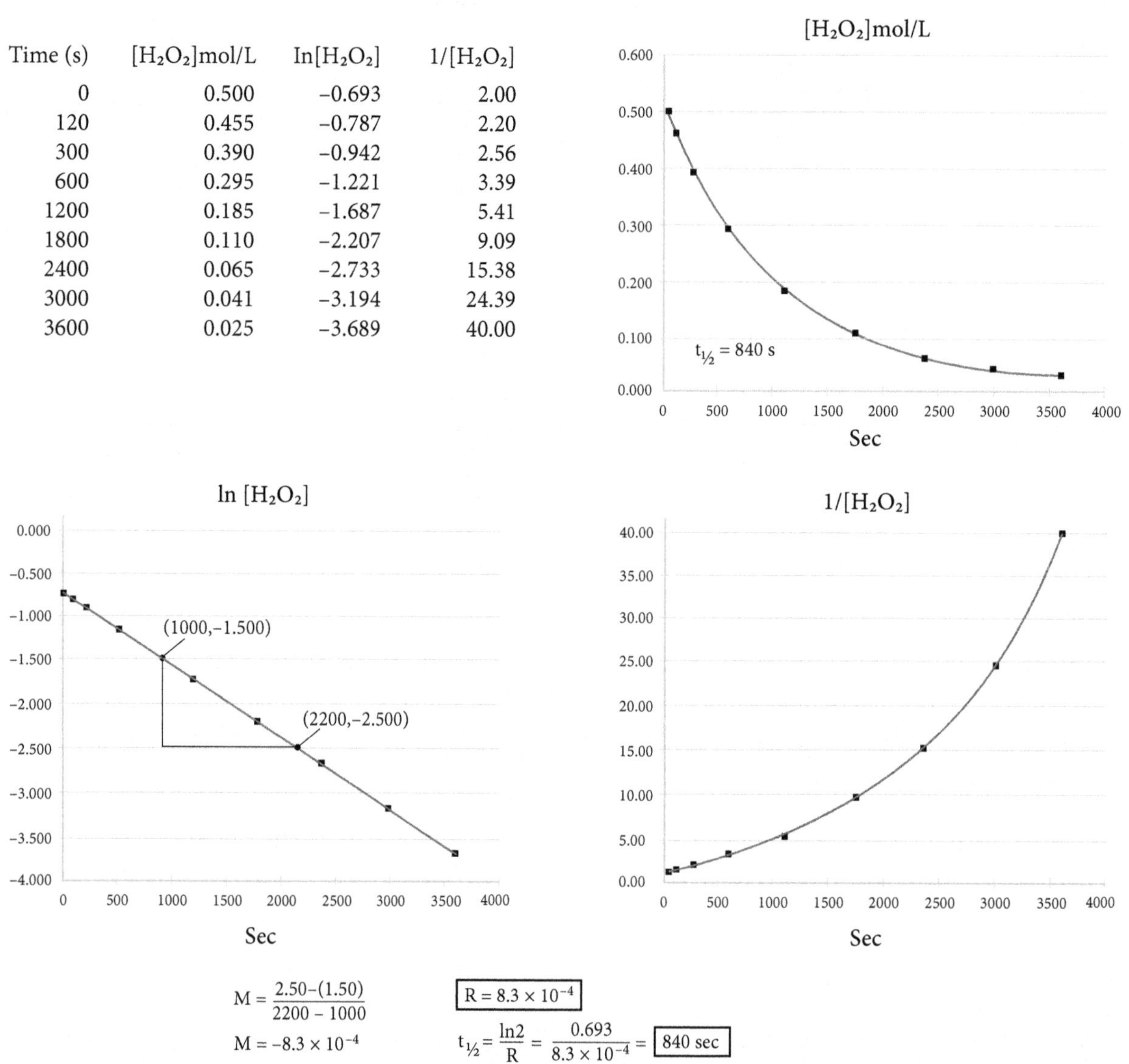

Time (s)	$[H_2O_2]$mol/L	In$[H_2O_2]$	$1/[H_2O_2]$
0	0.500	−0.693	2.00
120	0.455	−0.787	2.20
300	0.390	−0.942	2.56
600	0.295	−1.221	3.39
1200	0.185	−1.687	5.41
1800	0.110	−2.207	9.09
2400	0.065	−2.733	15.38
3000	0.041	−3.194	24.39
3600	0.025	−3.689	40.00

Figure 2.2 Decomposition of H_2O_2 Kinetics: Raw data (time, $[H_2O_2]$, and the linear plot of time (s) vs $\ln(H_2O_2)$; calculation of *k* and t½

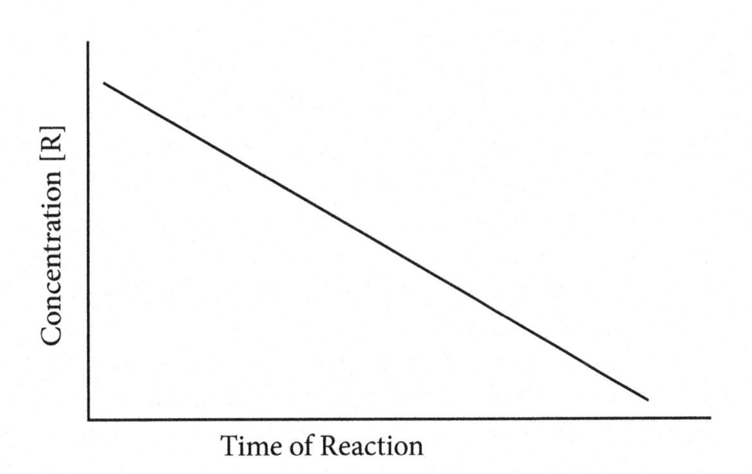

Figure 2.3 Plot of Concentration [R] versus time of reaction for a *zero-order reaction*

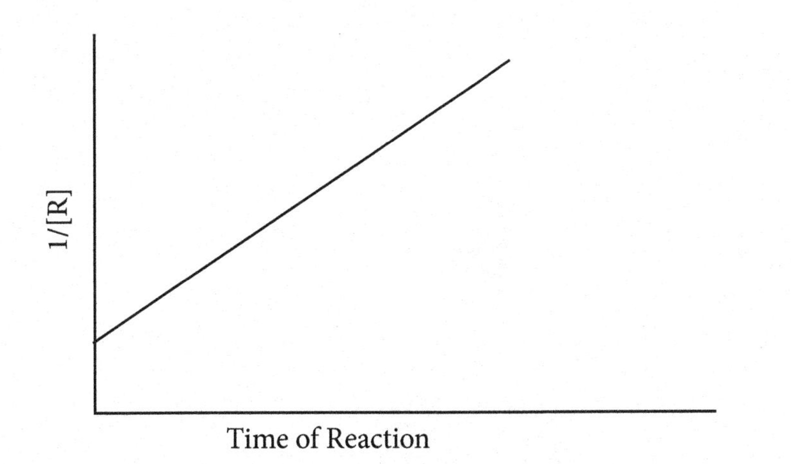

Figure 2.4 Plot of inverse concentration, 1/[R], versus time for a *second-order reaction*

What would be a real-world application of order of reaction? Suppose that there are a couple different reactions that produce the same chemical product but differ in their order of reaction. Thus, for a first-order reaction: doubling the concentration of reactant, [R] to [2R], would double the rate, $R_2 = 2R_1$. For a second-order reaction while doubling [R] to [2R] would quadruple the rate. If the goal is to quickly produce a desired product, then doubling [R] for a second-order reaction would result in a reaction that was four times faster than the original rate, $R_2 = 4R_1$. Chemists in Morocco studied a biodegradable plastic copolymer, which was degraded in two distinct reactions.[5] At the beginning of this degradation, the first reaction was found to be first-order, while the second one obeyed second-order kinetics. How did changes in concentration of the copolymer affect the rate of reaction? Doubling the amount of copolymer in the first reaction would result in doubling the rate, while in the second reaction doubling the amount would quadruple its rate. In both cases, the same reactant (i.e., the copolymer) is undergoing two different reactions with different rates of reaction.

Sec 2.3 Temperature, Kinetic Energy, and E_{act}

We can use progress of reaction diagrams (Figure 2.5 for an exothermic reaction) to explain how temperature affects the rate of a reaction. At a cooler reaction temperature, the molecules (atoms or ions) are moving slower as compared to their speed at a higher temperature. When a collision between chemical species occurs, chemical bonds may or may not break apart to form the "activated complex," which is found at the peak of the reaction curve. Specifically, if the kinetic energy (KE = ½ m * v^2) of the collision is <u>less than</u> the energy barrier, E_{act}, then the reaction will <u>not</u> occur. Conversely, if the KE is greater than the E_{act}, then the reaction <u>may occur</u> but it depends on other factors. A chemical reaction can result when the energetic collision (KE > E_{act}) has the *proper orientation* in space for the molecular species. Say that the "head" of one molecule must bump into the "tail" of another molecule for the reaction to occur. It follows that if the two heads collide with each other, then no reaction will occur. This is true even at elevated temperatures where the collisions possess enough energy (KE > E_{act}). Likewise, if the two tails collide, then no reaction will occur regardless of the temperature.

An example of how the *orientation of colliding molecules* affects reaction rate is as follows:

$$HCl\ (g) + H_2C{=}CH_2\ (g) \rightarrow H_3C - CH_2Cl\ (g)$$

This reaction occurs only when the collisions possess KE > E_{act} <u>and</u> the HCl molecule is properly oriented with the H end of HCl colliding with the C=C bond of $H_2C{=}CH_2$ as follows:

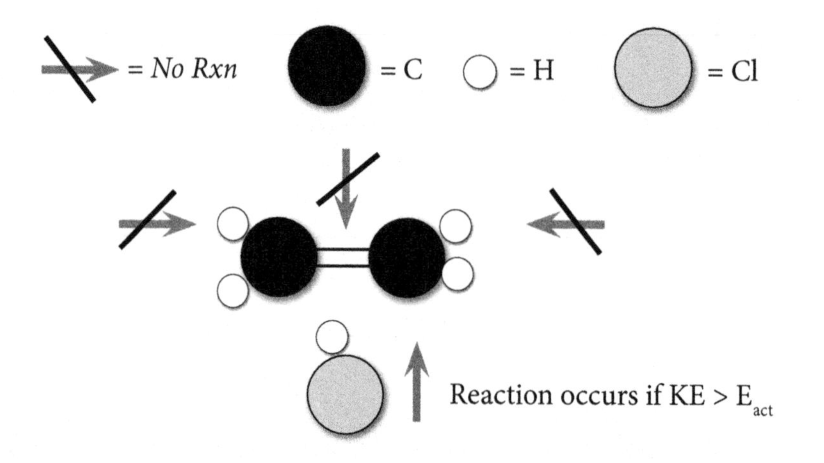

On the other hand, if the Cl end of HCl collides with the C=C double bond, <u>no</u> reaction occurs regardless of the kinetic energy of the collision. If the H end of HCl collides with the H_2C end of $H_2C{=}CH_2$, <u>no</u> reaction occurs. Thus, two factors affect whether or not a chemical reaction will occur: (1) the collision must provide enough energy to break a chemical bond (i.e., KE > E_{act}), and (2) the colliding molecules must have the proper orientation for the reaction to occur.

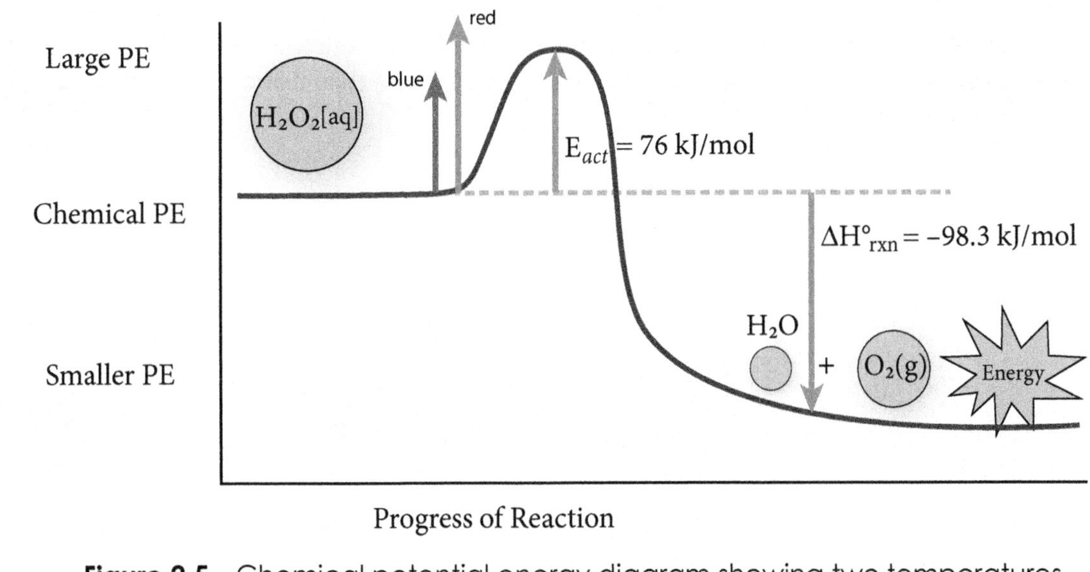

Figure 2.5 Chemical potential energy diagram showing two temperatures (blue = cooler temp ➔ less KE, and red = higher temp ➔ greater KE)

Let's now focus on the relationship between temperature and the KE of the collision with respect to the E_{act} of the reaction. One thing you should know is that the E_{act} is a constant for a given reaction regardless of temperature. So if a reaction is occurring at a lower temperature, it has the same E_{act} as it does at an elevated temperature. Specifically, in Figure 2.5 (above) the lower temperature is shown by the shorter "blue up arrow," while the elevated temperature is the longer "red up arrow." However, not all molecules at the same temperature are moving at the same speed (velocity). Especially for gas phase chemical reactions, there is a great range of molecular speeds from very slow to very fast. The best estimate of this phenomenon is the average kinetic energy of the molecules. The rule is that the *higher the temperature*, the *greater the average kinetic energy* of the molecules. Also, gas molecules are moving much, much faster than molecules in a liquid. As you will recall gas molecules move in rapid, random directions and speeds, and each molecule is independent of the velocity of nearby molecules. The much slower movement of molecules in liquids and solids is discussed in the textbook chapter on "intermolecular forces."

Sec 2.4 Colliding Molecules and Order of Reaction

How do colliding molecules (i.e., chemical species) relate to the "order of reaction"? For a second-order (i.e., bimolecular) reaction, two molecules collide with each other in the reaction. However, as described above, there are two factors that must be satisfied for a reaction to occur. First, these molecules must possess kinetic energy that is greater than the activation energy, E_{act}, for a particular reaction, and second, the molecules must have the proper orientation so they can react.

What about molecules in a first-order reaction? In a *unimolecular reaction*, only one molecule collides, but what does it collide with . . . ? A molecule can collide with the wall of its container, and if it has sufficient kinetic energy (i.e., KE > E_{act}) and proper orientation, then its bonds can be broken and hence a reaction can occur.

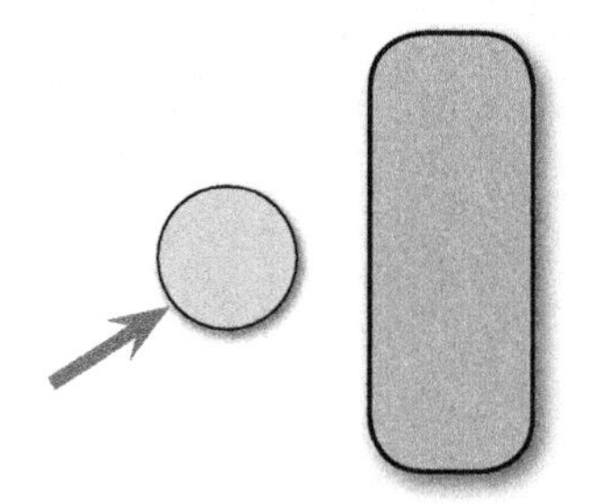

Furthermore, if the *unimolecular reaction* includes a catalyst (substance which speeds up a reaction), then the molecule can collide with the catalyst.

What about molecules in a zero-order reaction? For many reactions of this type, there is an excess amount of one of the reactants with respect to the limiting reactant. Thus, changing the amount of this excessive reactant does not affect the rate of reaction, and it is a zero-order reaction.

Sec 2.5 Effect of a Catalyst on E_{act}

A catalyst in a chemical system is a substance that increases the rate of a reaction without itself being consumed nor produced. Recall from reading your textbook that "half life" is the time required for the reaction to consume half of the original concentration of reactant. Specifically, for the H_2O_2 decomposition reaction, the half-life is 8½ days (e.g., [H_2O_2] from 0.90 M to 0.45 M). When an iron (III) catalyst is added to this reaction, the half-life is only 63 seconds.[6] What factor is responsible for this change in rate? The answer is the large difference in their energy of activation, E_{act} (See Figure 2.6).[7] For the catalyzed reaction (E_{act} = 56 kJ/mol) E_{act} is much, much lower than it is for the uncatalyzed reaction (E_{act} = 76 kJ/mol).[8] Thus, with this much lower energy barrier, the catalyzed reaction is much faster than the uncatalyzed reaction. Furthermore, when an enzyme is used as the catalyst, the activation energy drops even more: E_{act} = 7.1 kJ/mol.[9] Thus, the enzyme-catalyzed reaction is very rapid[9]. Catalase is one enzyme that decomposes H_2O_2 within living cells. In relatively high concentrations, hydrogen peroxide and its related oxidants can damage DNA.[10]

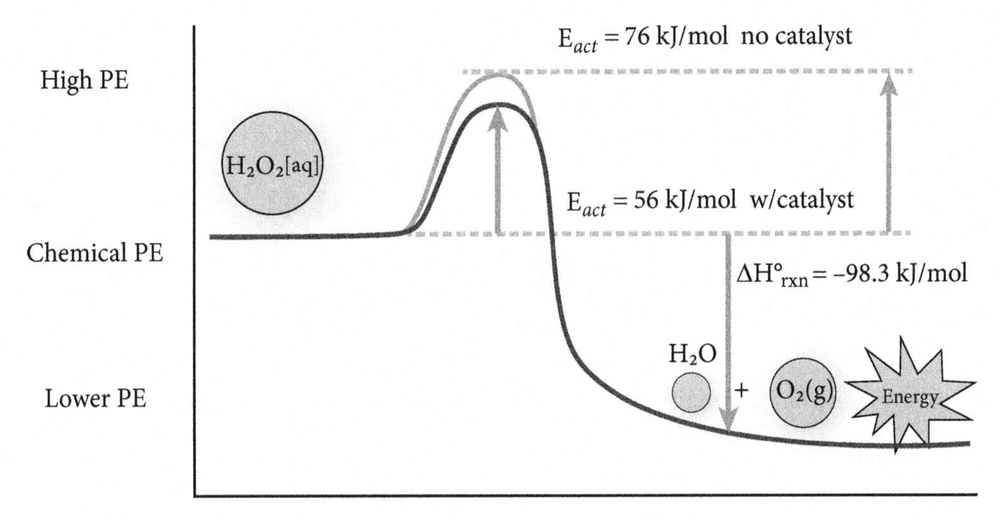

Figure 2.6 Chemical potential energy diagram showing two temperatures (blue = cooler temp ➔ less KE, and red = higher temp ➔ greater KE)

Sec 2.6 How Can We Compare Various Reaction Rates?

There are several ways to determine the *order of a reaction* for each of the reactants:

- Graph kinetics data to determine which plot is linear (see Sec 2.2);
- Perform several experiments where concentrations are systematically varied from one experiment to another:
  - First-order when doubling concentration [R], doubles the rate, $R_2 = 2R_1$
  - Second-order when doubling concentration [R], quadruples the rate, $R_2 = 4R_1$
  - Zero-order: doubling concentration [R], does <u>not</u> affect the rate, $R_2 = R_1$

In general, if the reaction is . . . A + B ➔ P ... then, ***Rate*** α $[A]^n * [B]^m$ where n and m are the respective orders for this reaction. There are two questions that arise: (1) What about the effects of temperature, nature of the reactants (e.g., Na versus Au reactivity), presence of a catalyst, and so on? (2) How can you calculate the *Rate* when a proportional sign, α, separates the two sides of the equation? The answer is that all of these factors are included in the rate constant, k, for the reaction. So if the *Rate* of a reaction is known for the concentration(s) of reactants, then the rate constant, k, can be determined as the only unknown in the rate law equation. Also, the rate constant can be calculated from the slope when a linear relationship is found. Specifically, when the plot of [A] versus time of reaction yields a line with a negative slope (equal to $-k$), the reaction is zero-order with respect to reactant A: $R = k[A]^o$ thus $R = k$. If the plot of natural log of concentration, ln[A], versus time is linear, then it is a first-order reaction with respect to reactant **A** and slope = $-k$: $R = k[A]^1$

Next, where the plot of inverse concentration, 1/[A], versus time is linear with a positive slope (slope = $+k$), then the reaction is second-order with respect to A: $R = k[A]^2$

Can the order of reaction be inferred from the balanced chemical equation? The answer is "no" because this must be determined from experiments. For example in this equation:

EQ 2.2: $F_2\ (g) + 2\ ClO_2\ (g) \rightarrow 2\ FClO_2\ (g)$

Experiments have shown that the rate law equation is first-order for both F_2 and ClO_2:

$$R = k\ [F_2]\ [ClO_2]$$

The *overall order of reaction* is equal to the sum of the orders for each reactant; where

$Rate = [A]^n * [B]^m$

The overall order is $n + m$. Thus, for the above reaction, EQ 2.2, the overall order is second-order (1 + 1 = 2).

Sec 2.7 Summary

Understanding the factors that affect the rate of a chemical reaction ultimately depends on understanding the progress of reaction diagram (e.g., Figures 2.5 and 2.6) for that particular reaction. The effect of temperature is easily explained as a comparison of the kinetic energy (KE) of the colliding molecules (i.e., chemical species) with the activation energy, E_{act}, for that reaction. If the KE of the collision is less than the E_{act}, then the reaction will not occur. If KE is greater than E_{act}, then the reaction may or may not occur—it depends on the second factor; that is, if the colliding molecules have the *proper orientation* as well as KE > E_{act}, then the reaction will occur. If an appropriate catalyst is used, then the reaction will be faster because the E_{act} has been lowered. Consequently, a catalyzed reaction can be just as fast at a lower temperature as it is at an elevated temperature. For example, in chemical industry, heating a chemical reaction requires lots of energy (and thus lots of money); whereas, using a catalyst (which is re-usable) with a lower temperature reduces the cost of running a particular reaction. In biological systems, an enzyme (biological catalyst) also serves this function, and it is highly specific for a particular reaction in the cell. Conversely, another factor—concentration of reactant—cannot be ascertained by looking at a progress of reaction diagram. Rather, one must plot the data as a graph of time-of-reaction (x-axis) versus concentration of reactant, [R] on the y-axis.

- If the plot is linear with a negative slope, then it is a *zero-order reaction*. This means that:
  - $R \propto [R]^0$, → doubling the concentration of R, results in the same rate
  - $R_1 \propto [1]^1$, $R_2 \propto [2]^0$, and $R_2 = R_1$
- If the plot is a curve with a negative slope, then it is either first-order or second-order reaction. To determine which one it is, you must transform the y-axis:
  - if the resulting natural log of the reactant, ln [R], versus time is linear with a negative slope, then it is a *first-order reaction*. This means that

- $R \alpha [R]^1$, ➔ doubling the concentration of R, doubles the rate
- $R_1 \alpha [1]^1$, $R_2 \alpha [2]^1$, and $R_2 = 2\, R_1$

❍ if the resulting inverse of reactant, 1/[R], is linear with a positive slope, then it is a second-order reaction. This means that

- $R \alpha [R]^2$, ➔ doubling the concentration of R, quadruples the rate
- $R_1 \alpha [1]^1$, $R_2 \alpha [2]^2$, and $R_2 = 4\, R_1$

Finally, you should understand the significance of the *rate law* for a given chemical reaction at a given temperature: For the generic reaction, A + B ➔ P, the rate law is written as $R = k\,[A]^m[B]^n$ where the exponents are determined by experimentation and/or graphical analysis. Remember that the rate constant, ***k***, is a very important component of the rate law because it reflects all factors other than [R] that affect the rate of reaction. So ***k*** represents the nature of the reactant (e.g., aqueous reaction of zinc versus gold), temperature ($T_2 = 10 * T_1$, then $k_2 = 2\, k_1$), with or without a catalyst ($k_{cat} > k_{no\text{-}cat}$), phase of the reaction ($k_{gas} > k_{liquid} > k_{solid}$), and particle size ($k_{dust} > k_{packed}$). Overall, we chemists hope you are O-*k* with all these factors. PS: Don't expect comedy-club humor from a chemist!

Sec 2.8 Technical References

Taken from free-access journals, and intended mostly for your chemistry professor's use.

[1] Brouwer, H. Small-Scale Experiments Involving Gas Evolution. *J. Chem. Educ.* **1995**, *72*(5), A100–A102.

[2] Abramovitch, D. A., Cunningham, L. K., & Litwer, M. R. Decomposition Kinetics of Hydrogen Peroxide: Novel Lab Experiments Employing Computer Technology. *J. Chem. Educ.* **2003**, *80*(7), 790–792.

[3] Justi, R., & Gilbert, J. K. History and philosophy of science through models: The case of chemical kinetics. *Science & Education*, **1999**, *8*(3), 287–307.

[4] Benarbia, A., Elidrissi, A., Ganetri, I., & Touzani, R. Synthesis, characterization and thermal degradation kinetics of copolyesters. *J. Mater. Environ. Sci.* **2014**, *5*(4), 8247.

[5] Haber, F., & Weiss, J. The catalytic decomposition of hydrogen peroxide by iron salts. *Proc. R. Soc. London, Ser. A.* **1934**, *147*, 332–351.

[6] De Laat, J. & Gallard, H. Catalytic decomposition of hydrogen peroxide by Fe(III) in homogeneous aqueous solution: mechanism and kinetic modeling. *Environ. Sci. Technol.* **1999**, *33*, 2726–2732.

[7] Tatsuoka, T., & Koga, N. J. Energy Diagram for the Catalytic Decomposition of Hydrogen Peroxide. *J. Chem. Educ.* **2013**, *90*(5), 633–636.

[8] Sweeney, W., Lee, J., Abid, N., & DeMeo, S. Efficient method for the determination of the activation energy of the iodide-catalyzed decomposition of hydrogen peroxide. *J. Chem. Educ.*, **2014**, *91* (8), pp 1216–1219.

[9] Bonnichsen, R. K., Chance, B., & Theorell, H. Catalase activity. *Acta Chem. Scand.*, **1947**, *1*, 685–709.

[10] Imlay, J. A., Chin, S. M., & Linn, S. Toxic DNA damage by hydrogen peroxide through the Fenton reaction in vivo and in vitro. *Science*, (29 Apr 1988) *240* (4852), 640–642.

The following references were not cited, but they indicate how students who have not studied this book have related to chemical kinetics.

[11] Cakmakci, G., Leach, J., & Donnelly, J. Students' ideas about reaction rate and its relationship with concentration or pressure. *Int'l. J. Sci. Educ.*, **2006**, *28*(15), 1795–1815.

[12] Cakmakci, G. Identifying alternative conceptions of chemical kinetics among secondary school and undergraduate students in Turkey. *J. Chem. Educ.*, **2010**, *87*(4), 449–455.

[13] Raviolo, A., & Garritz, A. Analogies in the teaching of chemical equilibrium: a synthesis/analysis of the literature. *Chem. Educ. Res. Pract.*, **2009**, *10*(1), 5–13.

[14] Van Driel, J. J. Students' corpuscular conceptions in the context of chemical equilibrium and chemical kinetics. *Chem. Educ. Res. Prac.*, **2002**, 201–213.

Sec 2.9 Questions to Ponder

Q1A. Given that two chemical reactions—one exothermic and the other endothermic—have the same activation energy, which one is more likely to "go to completion"?

1B. Which one would have the larger E_{act} for the reverse reaction?

1C. Which one would be the faster reverse reaction?

Q2A. Given that two similar reactions are conducted at the same temperature: Rxn A has a rate constant, k_A, that is twice that of Rxn B, k_B, and Rxn A is first-order while Rxn B is second-order, which one would be the faster reaction?

2B. Which reaction has the smaller E_{act}?

UNIT 3
What Is a Chemical System?

WHY STUDY THIS UNIT?

The goal of this unit is to help you develop a qualitative understanding of chemical systems. This is the concept that underlies almost all of the remaining topics covered in second-semester general chemistry. There is no corresponding chapter in any general chemistry textbooks.

Sec 3.1 Introduction

In first semester general chemistry, you learned about chemical equations, which you could balance and use to calculate the number of moles (or grams) that react with each other to form a certain number of moles (or grams) of each product. Also, you learned about chemical formulas and their relationship to moles and grams (via molar mass, *MM*). Each formula represents a chemical structure that has a particular type of bonding as well as a 3-dimensional shape (i.e., VSEPR shapes, such as linear, trigonal planar, tetrahedral, and so on). Furthermore, in this second semester course in general chemistry, you will use all of these concepts to learn about the different types of chemical systems.

A **chemical system** includes both the reactant(s) and product(s) as shown in its chemical equation; however, it also includes information used to predict if a chemical reaction will occur (this is called chemical thermodynamics) based upon the *chemical potential energies* of all of these chemical species (i.e., molecules/atoms/ions). As discussed in Unit 1, the law of conservation of energy applies to all chemical reactions, and it can also help you predict if a reaction is exothermic or endothermic. However, if you want to predict if a reaction will actually occur, then you need to also factor in the chemical species at the peak of the reaction diagram—where the most unstable species occurs (see Unit 2, chemical kinetics). As you will recall—a lower energy of activation, E_{act}, means a faster reaction than a similar reaction that has a higher E_{act}. So a chemical system also includes this "activated species" that is making the transition between the reactant and product chemical structures. If you want a reaction to react faster, then you can either increase the temperature or the concentrations of reactant(s), or lower the activation energy, E_{act}, of the reaction. It is easier to do the former but more challenging to do the latter. Recall that the proper catalyst for a chemical reaction speeds up the reaction by lowering its E_{act}. Any catalyst in a reaction is also part of its chemical system. In addition, the conditions under which the reaction occurs are part of the system. For example, temperature affects whether or not a reaction will occur, and thus it is part of the system. If a reaction occurs, then increasing the temperature by 10°C often results in the reaction being twice as fast. Pressure can affect the rate of reaction but only for certain gas phase reactions. Another factor is concentration of the reactant(s)—doubling the concentration can result in doubling the rate (first-order reaction), quadrupling the rate (second-order reaction—$2^2 = 4$X), or not affecting the rate (zero-order reaction). Overall, all of these conditions are needed to predict whether or not a reaction will occur, and if it occurs, whether it will be a "fast" or "slow" reaction.

Sec 3.2 What Are Interacting Chemical Species?

In second semester general chemistry, you are expected to begin to understand *how* a chemical reaction occurs, and to roughly predict *how much* of the reaction will occur (e.g., 10%, 50%, 90%, or 100%). These decisions are based on estimating the relative chemical potential energies of both the reactant(s) and product(s). For example, does ethane (C_2H_6) gas react with oxygen gas? If you draw circles to represent chemical potential energies, then you can predict: yes, the reaction will occur and, in fact, it will be exothermic.

EQ 3.1: $C_2H_6\,(g) + 7/2\,O_2\,(g) \rightarrow 2\,CO_2\,(g) + 3\,H_2O\,(l)$

C_2H_6 →

Does the balanced chemical equation, EQ 3.1, allow you to determine if it will react? No! It only shows *mass balance*. However, if you understand how the above circles represent the relative potential energies, then you can predict that the reaction will occur. Furthermore, by recalling the law of conservation of energy—you can predict that it is an exothermic reaction. That is, this method allows you to do an *energy balance* on this chemical reaction by showing heat energy as a "starburst":

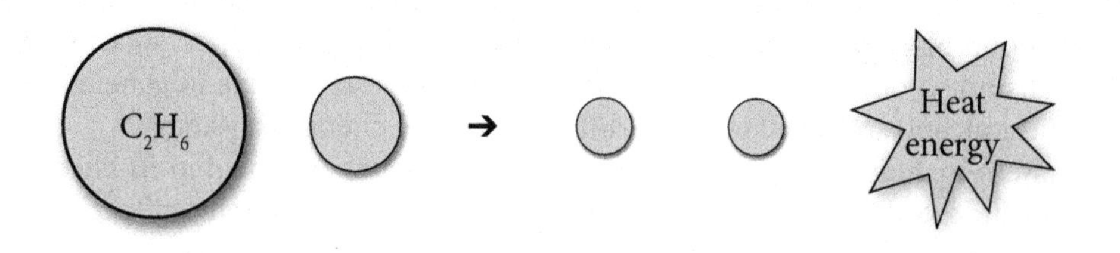

The above diagram could be modified to give quantitative information; however, you would need to look up the chemical potential energy (e.g., enthalpies of formation, ΔH_f) for each of these chemical species. These values are found in most textbooks in a table of thermodynamic values in the appendices. In your previous chemistry course, you probably recall calculating ΔH_{rxn} from the corresponding ΔH_f's for each compound in the equation. In this course, you will use and extend this type of quantitative determination in the textbook's chapter on thermodynamics (see Unit 7). For now, if you develop a *qualitative understanding* of energy balances, it will really help you learn all of the topics taught in this course (i.e., second semester general chemistry).

Sec 3.3 Dynamic Reactant/Product Relationships

In your previous chemistry course, all chemical reactions went to completion:

	R	→	P
Before rxn:	100%		0%
After rxn:	0%		100%

This reactant/product relationship is properly shown for many chemical reactions that "go to completion." This type of reaction is called an *irreversible reaction* because it only goes one way R ➔ P. The reason that they go to completion is that the chemical potential energies of the reactant(s) is/are much greater than the chem PE's of the products. In other words, the reactants are relatively unstable (reactive) and the products are stable (non-reactive). Any combustion reaction is an irreversible reaction because the enthalpy of reaction is very large—it gives off lots of heat—and carbon dioxide and water cannot reform the hydrocarbon (e.g., C_2H_6). The heat given off is very large: $\Delta H_{comb} = -1560$ kJ/mol.

EQ 3.1: $C_2H_6\ (g) + 7/2\ O_2\ (g) \rightarrow 2\ CO_2\ (g) + 3\ H_2O\ (l)$ + heat energy

The reason this reaction is "one way" is due to comparison of the E_{act} of the forward reaction with the E_{act} of the reverse reaction (Figure 3.1). Once the flame is ignited, the combustion of ethane supplies the heat energy to keep the flame going. This means E_{act} for the forward reaction is not too large—as compared to the E_{act} for the reverse reaction, which is much larger:

❍ *reverse* E_{act} = *forward* $E_{act} + |\Delta H_{rxn}|$ = *forward* E_{act} + 1560 kJ/mol (<u>not</u> –1560)

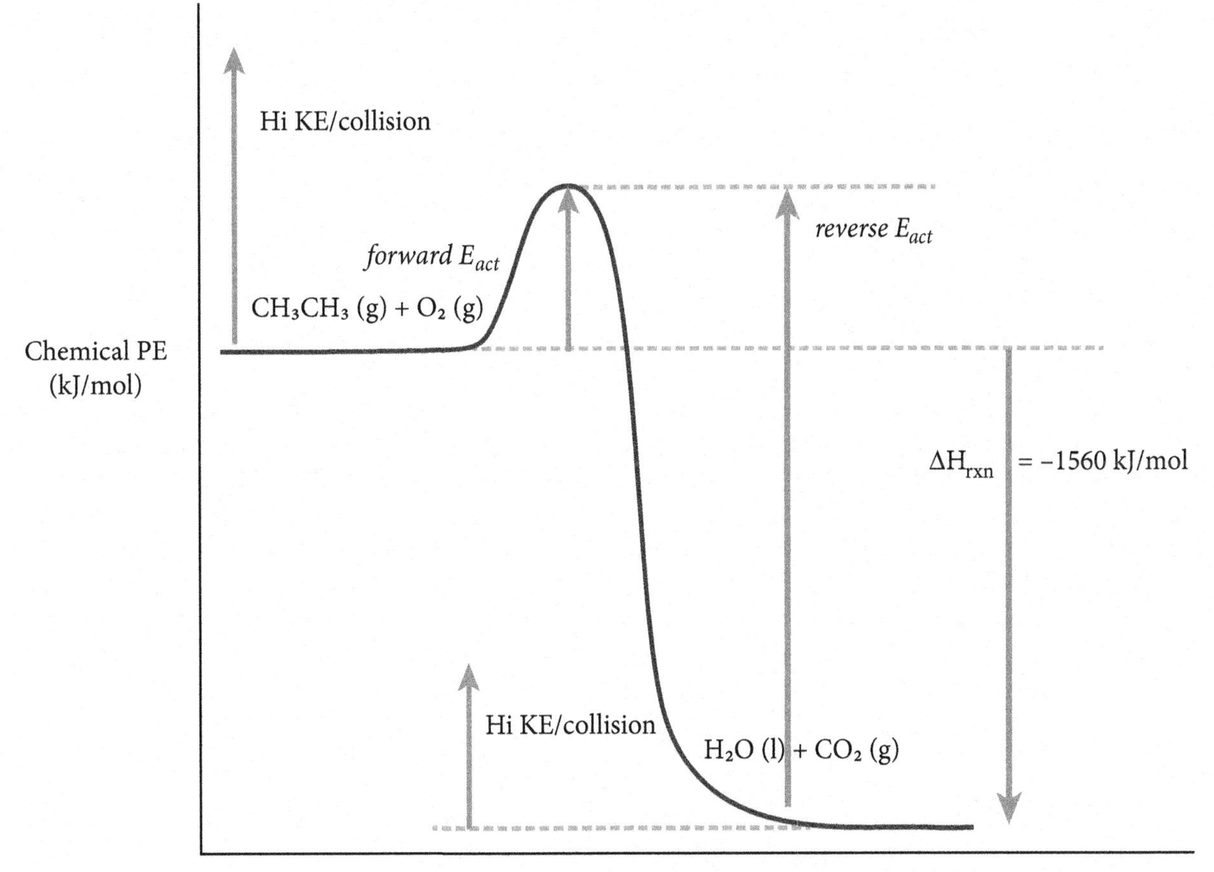

Figure 3.1 The reaction diagram for EQ 3.1

In Figure 3.1 (previous page), when an energetic collision between molecules occurs—as shown by the red arrow, "Hi KE/collision," the KE is greater than the *forward* E_{act}, thus, the reaction occurs as R ➔ P. However, when the same amount of kinetic energy of collision, "Hi KE/collision," between products, <u>no</u> P ➔ R reaction occurs because KE/collision is much less than the *reverse* E_{act}.

As opposed to the above reaction type, there is another one—one that does <u>not</u> go to completion because the product(s) can undergo a reaction to re-form the reactant(s). This type of reaction is called a *reversible reaction,* and the double-headed arrows are used to show that both reactions occur (R ➔ P and P ➔ R):

	R	$\leftrightarrow$	P
Before rxn:	100%		0%
After rxn:	??%		??%

Reversible reactions always have an experimental yield of product(s) that is far less than 100%. How can chemists determine the exact ratio of product(s) to reactant(s) for a particular reversible reaction? The answer is that they can predict which side will be predominant by using the relative chemical potential energies of the reactant(s) and products. When the reactant and product chemical PE's are nearly equal, it will be a reversible reaction. For example, when methyl acetate undergoes a reaction with water, the products are methanol, CH_3OH, and acetic acid, $HOCOCH_3$ as shown below:

EQ 3.2: $CH_3O{-}COCH_3\ (l) + H_2O\ (l) \leftrightarrow CH_3OH\ (l) + HO{-}COCH_3\ (l)$

Fruity odor --- mildly sweet vinegar-like smell

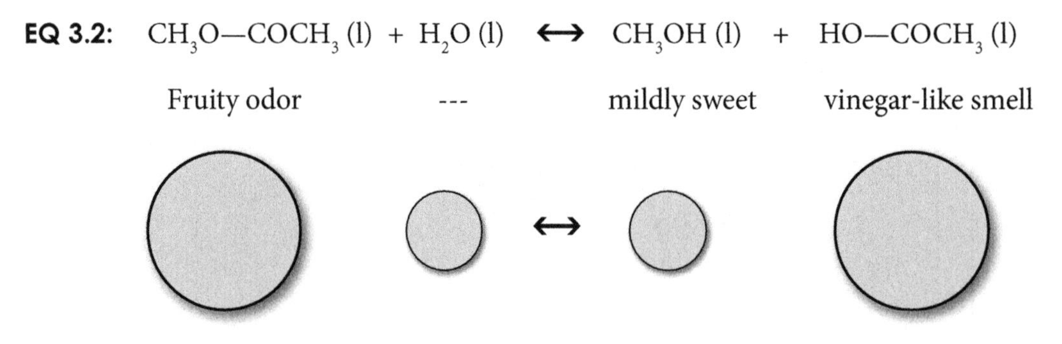

First, note that it is difficult to tell if this reaction is exo- or endo-thermic because each side has roughly equal chemical potential energies. When this is the case, the reaction is very likely to be a *reversible reaction* where both R ➔ P and P ➔ R occur simultaneously. For EQ 3.1, the actual enthalpy of reaction, ΔH_{rxn}, is only +6.79 kJ/mol. This means that the energy of activations for the forward and reverse reactions are nearly equal (see Figure 3.2 below). Thus, conditions that favor the forward reaction also favor the reverse reaction.

For the forward reaction, if the kinetic energy of an energetic collision is slightly higher than the forward E_{act}, then the R ➔ P reaction occurs. Likewise, when the same amount of KE occurs between product molecules, the P ➔ R reaction occurs. Thus, both reactions are occurring simultaneously.

$$Forward\ E_{act} = reverse\ E_{act} + \Delta H_{rxn} = reverse\ E_{act} + 6.79\ \text{kJ/mol}$$

Suppose that 0.10 M of each reactant reacts; meanwhile there are no product molecules. This situation is called *non-equilibrium* condition because the rate of forward reaction is much greater than the rate of the reverse reaction, which starts out at zero; that is, $\boldsymbol{R}_{forward} > \boldsymbol{R}_{reverse}$. When the two rates are <u>unequal</u>, then the reaction is at *non-equilibrium.* This process continues as the concentrations of reactants diminish and the concentrations of products increase. Eventually, the rate of the forward reaction becomes equal to the rate of the reverse reaction. This is the equilibrium condition, where

$R_{forward} = R_{reverse}$. Overall, these reversible reactions are the topic that will be studied in the next several chapters in your textbook.

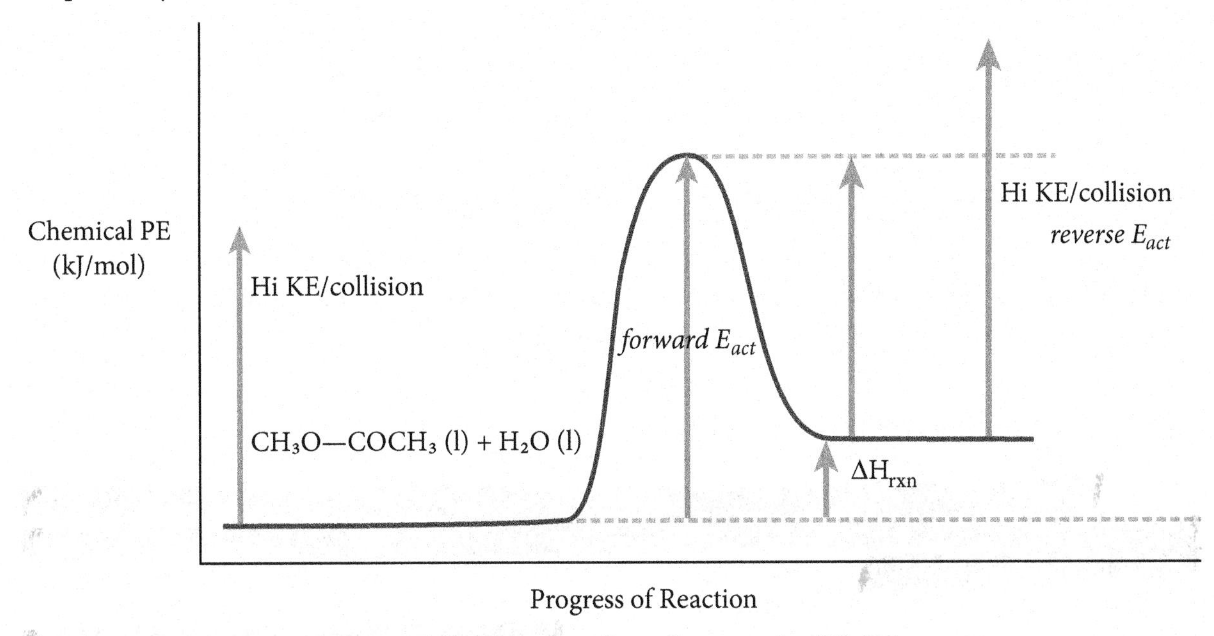

Figure 3.2 The reaction diagram for EQ 3.1

Sec 3.4 Physical/Chemical Condition

It is possible to have two similar chemical reactions that both produce the same product; however, the experimental conditions of the reactions can be quite different. The *physical/chemical condition* consists of the parameters (e.g., P, V, n, T) that govern the reaction and determine which product it will make. For example, ammonium chloride, NH_4Cl, can be produced under two different conditions:

EQ 3.3 $$HCl\ (aq) + NH_4OH\ (aq) \rightarrow NH_4Cl\ (aq) + H_2O\ (l)$$

This is an *aqueous phase reaction*, which initially begins at 25.0°C and results in a temperature of 31.8°C after the reaction has occurred. Is this reaction exo- or endo-thermic? The pressure before the reaction occurs is P = 1.00 atm and it remains the same throughout this reaction.

Under different experimental conditions, ammonia gas, NH_3 (g), reacts with hydrogen chloride gas, HCl (g), to produce ammonium chloride. In this *gas-phase reaction*, the initial pressure in the sealed reaction flask is P = 2.44 atm at 25°C, and after the reaction the final pressure is 0.244 atm at 25°C.

EQ 3.4 $$HCl\ (g) + NH_3\ (g) \rightarrow NH_4Cl\ (s)$$

Thus, this reaction features a large drop in pressure (because only a solid is produced) but no change in temperature. Conversely, the aqueous-phase reaction, EQ 3.3, features an increase in temperature with no change in pressure. Overall, these are two different reaction pathways to the same chemical species (i.e., NH_4Cl).

Sec 3.5 Predictable Manner of Chemical Behavior

Chemical processes respond in a predictable manner as described by the mathematics of exact relationships that accompany a change in one or more of the reaction parameters. During the remaining part of this course we will spend a lot of time addressing these mathematical relationships.

Sec 3.6 Changes in Reaction Parameters (Conditions)

Chemists can *manipulate* a chemical reaction to produce a desired product while diminishing the amount of an unwanted product. We will study how this "manipulation" occurs in several of the remaining chapters of your textbook. Hint: LeChatlier's Principle.

Sec 3.7 What Exactly Is a Chemical System?

Overall, a **chemical system** is a group of interacting chemical species that exist in a dynamic reactant/product relationship within a physical/chemical condition, which responds in a predictable manner to changes in that condition. *Interactive chemical species* are atoms, molecules, or ions that react with each other (see Sec 3.2). The basis for understanding these interactions is comparison of their relative chemical potential energies as discussed in Unit 1. The *dynamic reactant/product relationships* are determined by the extent to which the reaction goes to completion (see Sec 3.3). This type of process is where the reactant(s) completely form the products in an *irreversible process*, R ➔ P, which provides a 100% yield of product. Conversely, other reactions do not go to completion because the product forms the reactant in the reverse reaction (P ➔ R). This type of reaction is called a *reversible reaction* where both the forward and reverse reactions are occurring simultaneously, R ⟷ P. With this type of reaction, you never get a 100% yield of the product. The *physical/chemical condition* of a reaction is the set of experimental parameters (e.g., P, V, n, T) that describes to a chemist the "recipe" for how to set up an experiment in order to bring about the desired chemical reaction (see Sec 3.4). Chemical processes respond in a *predictable manner* based on mathematical expressions, which describe the exact relationship among the reaction parameters as a result of the chemical reaction (see Sec 3.5). Finally, a chemist can manipulate a chemical reaction by changing the reaction parameters such that the desired product(s) are made in a higher yield than would be possible in the absence of this manipulation (see Sec 3.6).

UNIT 4
Equilibrium in Chemical Systems

WHY STUDY THIS UNIT?

The goal of this unit is to help you understand "chemical equilibrium" as a feature of reversible reactions (where R ➔ P and P ➔ R both occur). In this unit most of the chemical reactions occur in the gas phase, while in Unit 5 all reactions are conducted in aqueous solutions. Chemical equilibrium is covered in Chapter 13, 14, or 15 of your general chemistry textbook.

Sec 4.1 Introduction

Before you begin your study of chemical equilibrium, you need to be aware that there are two distinct types of chemical reactions. *Irreversible reactions* tend to go to completion (e.g., R ➔ **P** where there is 100% **P** at the end of the reaction). A common example of an irreversible reaction is the explosive mixture of hydrogen gas and oxygen gas, which just needs a spark of energy to ignite this exothermic reaction (See Figure 1.4A).

EQ 4.1: $2\,H_2\,(g) + O_2\,(g) \rightarrow 2\,H_2O\,(l) + \text{energy}$

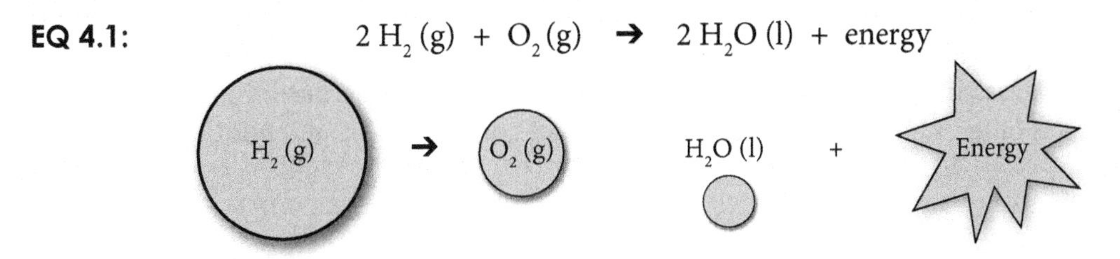

When stoichiometric amounts of the reactants (2:1 mole ratio of H_2 to O_2) are ignited, this reaction occurs in less than a second to produce liquid water and energy (i.e., exothermic reaction). Assuming an exact 2:1 reaction ratio, if you analyzed the product mixture, you would find 100.0% water and 0.0% hydrogen and oxygen gases. Thus, we say that this reaction "goes to completion."

You may wonder how an endothermic reaction can occur—especially since it is often called a *nonspontaneous reaction.* This type of reaction requires a continuous input of energy in order to make the reaction occur. For example, let's look at the backward reaction for EQ 4.1, which is shown below in EQ 4.2:

EQ 4.2: $2\,H_2O\,(l) + \text{energy} \rightarrow 2\,H_2\,(g) + O_2\,(g)$

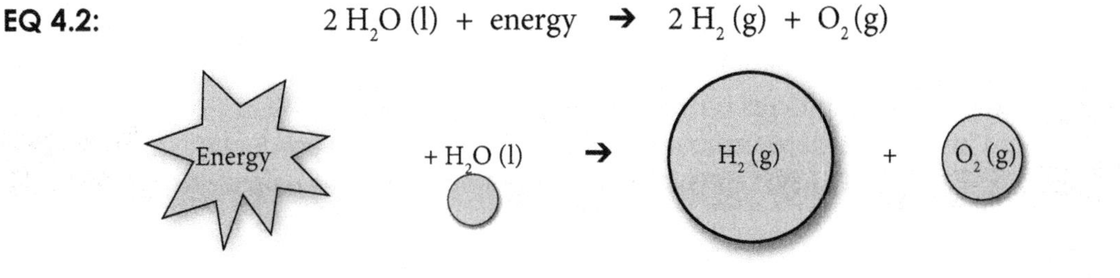

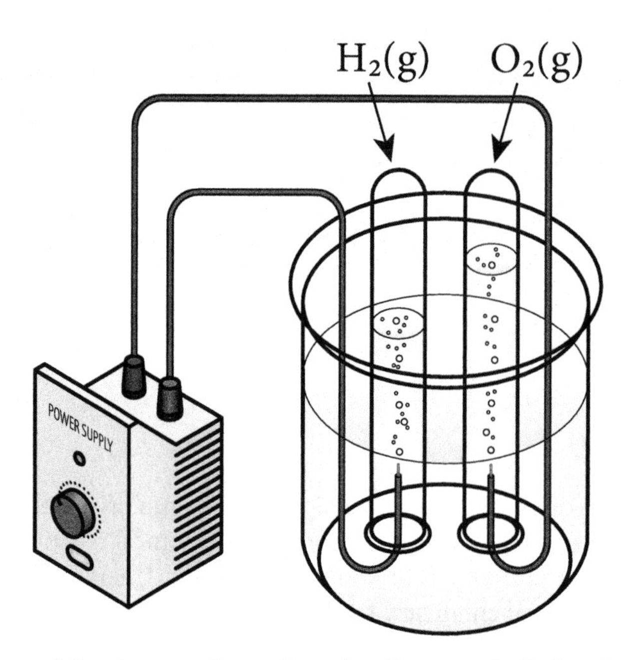

Figure 4.1 Apparatus set up for the electrolysis of water
©Zern Liew/Shutterstock.com

For this backward reaction: Why do hydrogen and oxygen gases not spontaneously react to form water? The answer is that when the chemical reaction shown in EQ 4.2 is done in the laboratory, the two gaseous products are completely separated from each other as shown in Figure 4.1 above. From the stoichiometric ratio shown in this equation, you should be able to tell that the tube on the left is collecting H_2 (g) while the one on the right (half the volume) is collecting O_2 (g). Thus, these two highly reactive gases in this reaction cannot react with each other because they are completely separated from each other. Overall, it usually takes unusual conditions for an endothermic reaction to occur.

Sec 4.2 Reversible Reactions and Equilibrium

As described above, irreversible reactions, R → P, are reactions that produce only the product(s). On the other hand, *reversible reactions*, R ⟷ P, produce a mixture of the reactant(s) and product(s). At the very beginning of a chemical reaction, it is difficult to distinguish between these two types of reaction because only reactant(s) is/are present, and hence only R → P can occur. Eventually, a distinction occurs for a *reversible reaction,* due to the fact that as the product accumulates, the reverse reaction begins to consume the product, P → R. During this first stage, the rates of the R → P and P → R reactions are different, which is called *non-equilibrium.* That is, the two rates are unequal: the rate of the forward reaction, $R_{forward}$ (R → P) is faster (or slower) than the rate of the backwards reaction, $R_{backwards}$ (P → R). In the second stage, the faster one slows down and the slower one speeds up until their rates are equal: $R_{forward} = R_{backwards}$. This process is called *equilibrium.*

Does "equilibrium" mean that there are equal amounts of reactants and products? No! Equilibrium means that for every time a reactant molecule (ion or atom) forms a product molecule, a product molecule reacts to form a reactant molecule. A chemist or chemistry student can use any amount of reactant to start the reaction: [R], where [ ] means concentration in *molarity*, number of moles per liter. Next, for a reversible reaction, you can start with any molarity of reactant and any molarity of product. The question is . . . How will the reversible reaction respond to these variable amounts of reactants and products? The answer to this question is usually shown in your textbook as a table that shows several different experiments that differ in the initial amounts of reactants and products. Typically, the first several experiments have initial amounts of reactant but zero concentrations of product, then vice versa (see Table 4.1). For example, a very common chemical system that illustrates these experiments is the NO_2–N_2O_4 system[1 & 2].

EQ 4.3: $N_2O_4\ (g) \leftrightarrow 2\ NO_2\ (g)$ at 100°C

In Experiment 1, the initial concentrations of $[N_2O_4] = 0.0$ and $[NO_2] = 0.0200$ M, and when the reaction attains equilibrium, the resulting K_c is a constant, which is calculated from the equilibrium concentrations. The question is . . . was this K_c just a simple ratio of concentrations at equilibrium? And the answer is "no" because for Experiment 1, $[NO_2]_{eq}$ to $[N_2O_4]_{eq}$ ratio is equal to [0.0172]/[0.00140] = 12.3, while for Experiment 2, this ratio is equal to [0.0308]/[0.00451] = 6.83. Thus, the simple equilibrium ratio is not constant. Chemists then tried to use the balanced chemical equation (see EQ 4.3), and recalculate the ratio as follows:

$$K_c = [NO_2]^2_{eq}/[N_2O_4]_{eq}$$

The resulting calculation yields a constant value regardless of what set of initial concentrations are used to calculate K_c.

Expt 1: $K_c = [0.0171]^2/[0.00140] = 0.209$

Expt 2: $K_c = [0.0308]^2/[0.00451] = 0.211$

You probably recall that when you do a chemistry experiment, there is always a degree of uncertainty in the right-most digit. Thus, for chemists: $K_c = 0.209$ and $K_c = 0.211$ gives an average of 0.210 with an error of only 0.48%. Chemists generally believe that any experiment where results are within 1% of the expected value, represents excellent results.

Table 4.1 Initial and Equilibrium Concentrations of $[N_2O_4]$ and $[NO_2]$ at 100°C

Expt	$[N_2O_4]_{init}$	$[NO_2]_{init}$	$[N_2O_4]_{equil}$	$[NO_2]_{equil}$	K_c
1	0.0	0.0200	0.00140	0.0171	0.209
2	0.0200	0.0	0.00451	0.0308	0.211

Overall, the generic expression for a chemical reaction can be shown as follows, where A, B, and AB are chemical species (molecules/ions/atoms) and *x, y,* and *z* are the coefficients in the balanced equation.

$$x\,A\ (g) + y\,B\ (g) \leftrightarrow z\,AB\ (g)$$

Next, the generic expression for the equilibrium constant, K_c, is as follows:

$$K_c = [AB]^z /([A]^x * [B]^y)$$

You can read your textbook to get other examples that show how K_c is a constant for any reversible reaction that has attained equilibrium at a particular temperature.

Sec 4.3 Reversible Reactions and the Meaning of K_c

Equilibrium constants for different reactions can vary greatly in the range of their values. For example, a chemical reaction that can remove sulfur from the combustion of sulfur-laden coal (see EQ 4.5 in Sec 4.4, below) has an equilibrium constant, $K_c = 8.0 * 10^{15}$, which is much, much greater than one. If expressed in US dollars, what would the exponent of 10^{15} mean? Suppose that one million rich people (i.e., billionaires) are gathered together—the sum of all their money put together would be \$1,000,000,000,000,000 (10^{15}). In terms of number of molecules found at equilibrium, it means that for every reactant molecule you find, there would be one million-billion product molecules ($1 * 10^{15}$). Thus, this reaction is essentially "going to completion."

At the other extreme, some reversible reactions have equilibrium constants that are much, much smaller than one. For example, the decomposition of hydrogen bromide gas into its elements in their gaseous state has K_c equal to $7.7 * 10^{-11}$. How small is this number? What if you needed to locate another person who could be anywhere on planet Earth? What if you had absolutely no idea where they are located? If you were just guessing where the person is located, your odds of finding that person on planet Earth would be 1 out of $1.4 * 10^{10}$ (14,000,000,000 is the population of Earth). That is, probability = $1.4 * 10^{-10}$ of randomly finding that person. Suppose you got very lucky and found out that the person was now visiting Inner Mongolia, CHINA—in the city of Kweisui (Hohhot or "Blue City"). There are 1.4 million people living in this city—so now your odds are much better—roughly one in a million! Overall, these examples show that these numbers are very, very small.

When compared to the above extremes among the values of K_c, we can now discuss numbers that have moderate values that are roughly "close to one." Chemists generally put reversible reactions in this moderate category when the K_c is a ratio between 0.001 (10^{-3}) and 1,000 (10^3) of product to reactant molecules. To continue with the above analogy, if you know the person is somewhere in a town of 1,000 people, then you have much better odds of finding them as compared to the odds of finding them somewhere on Earth.

Sec 4.4 Reversible Reactions and Moderate Values of K_c

When a reversible reaction has a K_c in the moderate category, $10^{-3} < K_c < 10^3$, the number of molecules (ions/atoms) on the reactant and product sides are roughly equal to each other. In general:

Given:	A_2 (g)	+	B_2 (g)	$\leftrightarrow$	2 AB (g), then
#molecules at equilibrium	6		6		6

$K_c = [AB]^2/[A_2]*[B_2]$

$K_c = [6]^2 /[6]*[6] = 1.0$

Is it normal for all of the equilibrium concentrations to be equal to each other? The answer is "no"! For example, look at the reaction below:

EQ 4.3: N_2O_4 (g) ⟷ NO_2 (g) at 127°C $K_c = 1.44$ (from experiments)

#molecules			
Expt 1:	6	6	$K_c = 6.0$ {No, wrong ratio → non-equilibrium}
Expt 2:	17	5	$K_c = 1.47$ {proper ratio → equilibrium }
Expt 3:	35	7	$K_c = 1.40$ {proper ratio → equilibrium }

$K_c = [NO_2]^2_{eq}/[N_2O_4]_{eq}$

$K_c = [5]^2_{eq}/[17]_{eq} = 1.47$ (which is very similar to the true value, $K_c = 1.44$)

Thus, as shown above for Experiment 2, if you have $[NO_2]_{eq} = 5$ molecules and $[N_2O_4]_{eq} = 17$ molecules, this system would be at equilibrium! Also, for Experiment 3, $K_c = 1.40$, and thus 35 molecules of N_2O_4 and 7 molecules of NO_2 constitutes an equilibrium mixture of these gases. In terms of "counting molecules," these "experiments" can be illustrated by the example shown below:

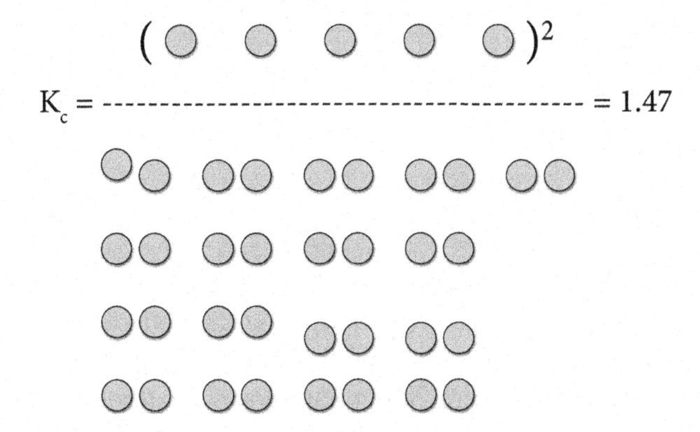

Overall, if the calculated ratio for any set of concentrations produces an equilibrium constant equal to $K_c = 1.44$, then that system is at equilibrium for the reaction in EQ 4.3. Does temperature influence this equilibrium ratio? Yes, the value of K_c changes with the reaction temperature.

Another example of a reversible reaction that yields an equilibrium constant in the moderate range (i.e., Kc = 10^{-3} to 10^3), is sulfuryl chloride, SO_2Cl_2. Sulfuryl chloride is used in the industrial production of *pesticides*, and to treat wool to prevent shrinking. The equilibrium constant, K_c, for this reaction is roughly equal to one; thus, the equilibrium mixture contains very similar amounts of reactant and products:

EQ 4.4: SO_2Cl_2 (g) ⟷ SO_2 (g) + Cl_2 (g) $K_c = 2.39$ T = 90°C

Thus, we can write the equilibrium expression as follows:

$$K_c = \frac{[SO_2]_{eq} * [Cl_2]_{eq}}{[SO_2Cl_2]_{eq}}$$

Why does this reaction have a K_c in the moderate range? The answer is that the E_{act} for both the forward reaction and the backward reaction are similar to each other. This reaction is endothermic, but its value is relatively small: ΔH_{rxn} = +58 kJ/mol. This smaller value for the enthalpy change of a reaction means both the forward and reverse reactions are occurring at about the same extent. When equilibrium, $R_{forward} = R_{reverse}$, is achieved, there are roughly about the same amounts of product(s) and reactant(s).

		SO_2Cl_2 (g)	↔ SO_2 (g)	+ Cl_2 (g)	K_c = 2.39 T = 90°C
#molecules	Expt 1:	11	5	5	K_c = 2.3 {proper ratio}
	Expt 2:	5	5	5	K_c = 5.0 {non-equilibrium}
	Expt 3:	12	4	7	K_c = 2.3 {proper ratio}

So, when looking at the proper counting of molecules as shown in Experiment 1, K_c is equal to:

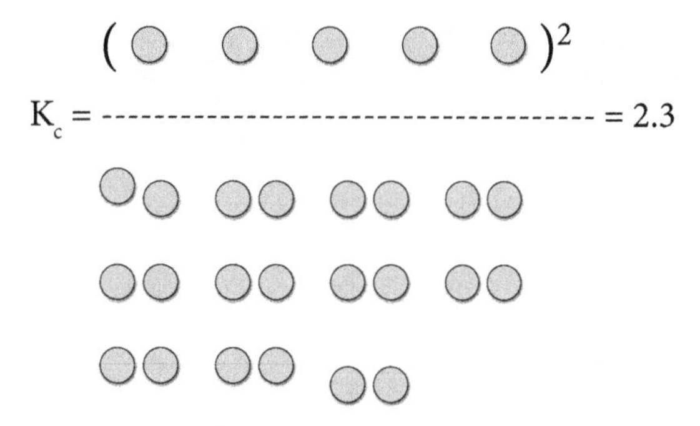

Please note that in the case of equal numbers of product molecules for both SO_2 and Cl_2, that we can multiple them together: [5] * [5] =25, or square them: $[5]^2$ = 25. Overall, any particular ratio that gives an equilibrium constant, K_c = 2.3, constitutes a "proper count" of molecules at equilibrium for the reaction in EQ 4.4. So Experiment 3 is also a system at equilibrium. What happens if the temperature of this reaction changes? Then the value of the equilibrium constant must be recalculated based on the resulting amounts of reactant(s) and product(s) established at equilibrium.

Sec 4.5 Reversible Reactions and Very Large Values of K_c

Some chemical reactions at equilibrium yield a huge count of product molecules, but a very, very small count of reactant molecules. These "reversible reactions" have a very, very large value for their equilibrium constant (i.e., $K_c >> 1$). One example of this type of reversible reaction is shown below. This reaction is used to remove sulfur dioxide gas, SO_2, from smoke stacks. Sulfur dioxide is a pollutant that produces "acid rain" when it combines with water to make sulfuric acid, H_2SO_4 (aq). This type of rain has a very low pH (acidic pH) that damages trees and other green plants as well as harming aquatic animals (e.g., frogs and fish). Please note that the equilibrium constant for this reaction is very large: $K_c = 8.0 * 10^{15}$. This means that at equilibrium, there are millions and millions of product molecules for every single molecule of reactant. Also, this reaction occurs at room temperature (i.e., 25°C).

EQ 4.5: SO_2 (g) + 2 H_2S (g) ↔ 3 S(s) + 2 H_2O (g) $K_c = 8.0 * 10^{15}$ T = 25°C

This reaction is exothermic $\Delta H_{rxn} = -147$ kJ/mol

Do you notice anything different when you look at the products of this reaction? Sulfur, one of the products, is a yellow solid at room temperature. You should recall from reading your textbook that this type of reaction is called a *heterogeneous reaction*. In a gas phase reaction, only the gaseous reactants and products are included in the K_c expression. All solid and liquids are ignored. Why? The solid, yellow sulfur in EQ 4.5 must be present when equilibrium is reached; however, the amount of solid does not affect the K_c value.

How then should a solid or liquid be represented? Answer: It is best to think of the amount of solid as being equal to one: [S(s)] = 1; thus $[1]^3 = 1$ and this number does not affect the value of the equilibrium constant. So we ignore sulfur in the K_c expression:

$$K_C = \frac{[H_2O]^2_{eq}}{[SO_2]_{eq}\,[H_2S]^2_{eq}}$$

$$SO_2\,(g) + 2\,H_2S\,(g) \longleftrightarrow 3\,S(s) + 2\,H_2O\,(g) \quad K_c = 8.0 * 10^{15}\ T = 25°C$$

Suppose that there is one molecule of SO_2 and one molecule of H_2S at equilibrium. How many molecules of H_2O (g) would there be at equilibrium? Roughly speaking, there would be about 10 million H_2O (g) molecules. Thus, $[H_2O]_{eq} = 1 * 10^7$ molecules, then $[H_2O]^2 = 1 * 10^{14}$; however, more accurate estimate would be . . . if $[H_2O]_{eq} = 9 * 10^7$ molecules, then $K_c = 8.1 * 10^{15}$

So actually there would be 90 million molecules of steam, H_2O (g), for every molecule of SO_2 and H_2S. How do chemists explain this huge ratio of products to reactants? The answer depends on the relative *chemical potential energy* (Unit 1). Both SO_2 and H_2S are highly reactive molecules—SO_2 is a pollutant that can produce acid rain and H_2S is a deadly toxic gas. Hydrogen sulfide disrupts cellular metabolism in much the same way that cyanide (HCN) leads to cell death[3]. If you ever have your nose above a H_2S generator {e.g., Na_2S (aq) + HCl (aq) → H_2S (g) + 2 NaCl (aq)}, then are you alive? As long as you can smell the "rotten egg" smell of H_2S, you are "okay." However, the moment that you can no longer detect it, it is probably too late. Therefore, <u>never</u> acidify any Na_2S solution (or any M_2S solution, in general) and never stay around an H_2S generator for any length of time.

What factor could cause a reversible reaction to have a K_c that is much, much larger than one? In general, when a reactant in the balanced chemical equation is very reactive (i.e., unstable molecule), and the product(s) are relatively stable. Recall that unstable molecules are represented with very large circles (i.e., they have very high chemical potential energy), while stable molecules are shown as very small circles (much lower chemical PE).

Suppose that the chemical reaction shown in EQ 4.5 was conducted under different experimental conditions (e.g., different temperature) and the "new" equilibrium constant was found to be $K_c = 9.8 * 10^3$. What plausible amounts of SO_2, H_2S, and H_2O gases at equilibrium ($R_{forward} = R_{reverse}$) would result in this value for K_c? As shown below, if the mixture was $[H_2O]$ = 99 molecules, $[SO_2]$ = 1 molecule, and $[H_2S]$ 1 molecule, then it would be at equilibrium because: $K_c = [99]^2 / (1*1) = 9801 = 9.8 * 10^3$ at the given temperature. On the other hand, what if that temperature was 25°C? This means that $K_c = 8.0 * 10^{15}$ and we would need to draw 10 million molecules of H_2O gas, 1 molecule of SO_2, and 1 of H_2S.

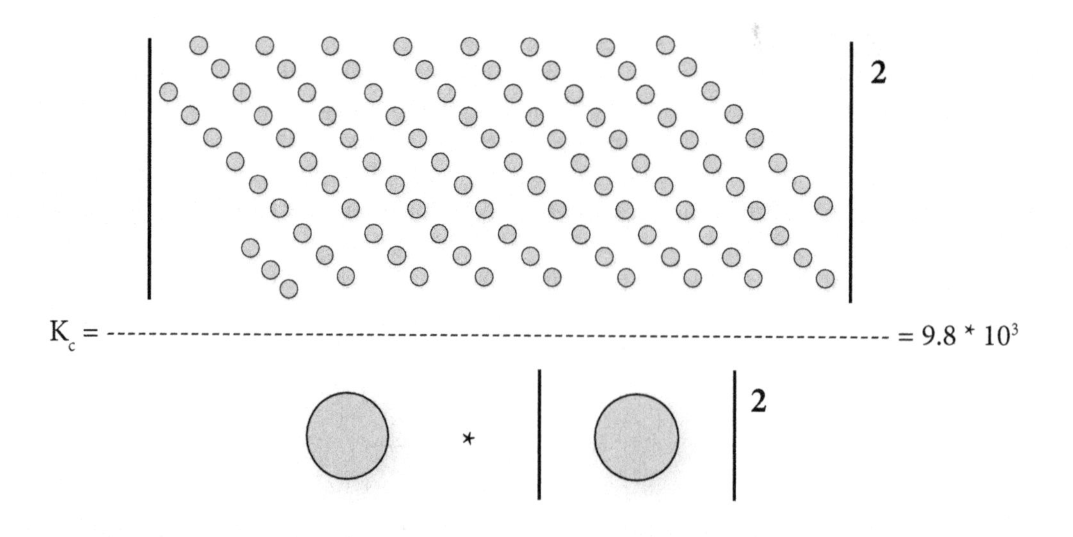

Overall, when you read your textbook, remember that there are three ways that you can count molecules: concentration in molarity (e.g., 0.60 M = [0.60]), partial pressure of gases (e.g., P_{gas} = 0.60 atm), or counting molecules (as shown above). Of course molarity and partial pressures are indirect methods of counting molecules; however, these amounts can be determined by experiments with reversible reactions.

What if water is written as a liquid in EQ 4.5? Then the resulting equilibrium expression would be written as:

$$K_c = \frac{1}{[SO_2]_{eq}\,[H_2S]^2_{eq}} \qquad \text{or} \qquad K_c = ([SO_2]_{eq}\,[H_2S]^2_{eq})^{-1}$$

Thus, you need to always pay attention to the physical state (s, l, or g) of each chemical species in a reversible reaction.

Sec 4.6 Reversible Reactions and Very Small Values of K_c

When the equilibrium constant for a reversible reaction is much, much lower than one (K_c << 1), what does this mean? When hydrogen chloride gas decomposes to form hydrogen and chlorine gases at 27°C (300 K), the equilibrium constant is very, very small: $K_c = 2.0 * 10^{-9}$.

$$2\,HCl\,(g) \leftrightarrow H_2\,(g) + Cl_2\,(g) \qquad K_c = 2.0 * 10^{-9} \text{ at } T = 27°C\ (300\ K)$$

It would be very had to visualize this small number. However, if this reversible reaction was found to occur at a different temperature, 727°C (1000 K) where $K_c = 1.0 * 10^{-4}$, then there would be 99 molecules of HCl (g) and 1 molecule of H_2 with 1 molecule of Cl_2. Please note that— $(99)^2 = 9801$, and $K_c = 1/9801 = 1.0 * 10^{-4}$

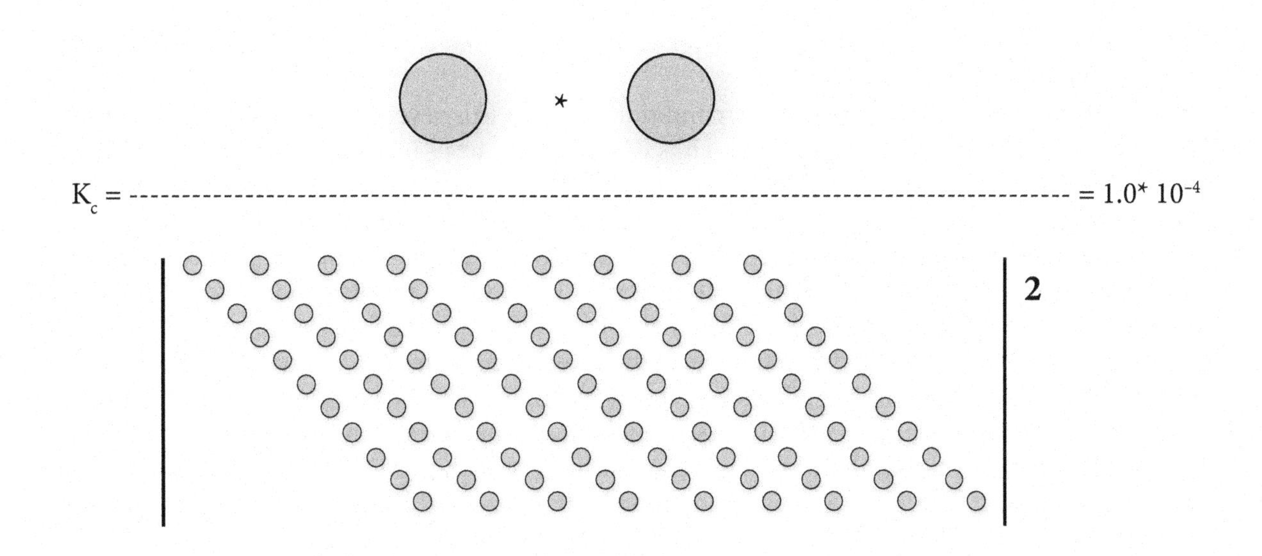

Sec 4.7 Reversible Reactions and Non-Equilibrium

At the beginning of most reversible reactions—the system is usually not at equilibrium. This is because when chemists are planning how to do a chemical reaction in the lab, they normally start with 100% reactants. Thus, there is a huge excess of reactants, and the forward reaction is the only possible reaction (because [Prod] = 0). Then just after the beginning of the reaction, the ratio of product molecules to reactant molecules is probably <u>not</u> equal to the equilibrium amounts of reactants and products. For example, the decomposition of SO_2Cl_2 (refer back to EQ 4.4) has an equilibrium constant equal to 2.39.

EQ 4.4: $SO_2Cl_2\ (g) \longleftrightarrow SO_2\ (g) + Cl_2\ (g) \qquad K_c = 2.39 \qquad T = 90°C$

Suppose there are 99 molecules of SO_2Cl_2, 1 molecule of SO_2, and 1 molecule of Cl_2. We can then write the non-equilibrium expression, Q_c, as follows:

$$Q_c = \frac{[SO_2]_{\text{non-eq}} * [Cl_2]_{\text{non-eq}}}{[SO_2Cl_2]_{\text{non-eq}}}$$

Next, we can calculate the value of Q_c from the given number of molecules (99 of SO_2Cl_2, 1 of SO_2, and 1 of Cl_2), and then compare this answer to the value of K_c.

$$Q_c = \{[1] * [1]\}/[99] = 0.010$$

Please note that both K_c and Q_c are ratios of products to reactants, where the balanced chemical equation is used to provide the appropriate exponent for each chemical species (i.e., molecules). Thus, when Q_c is greater than K_c, there are too many product molecules with respect to the number of reactant molecules. Likewise, when $Q_c < K_c$, then there are not enough product molecules with respect to reactant molecules. Finally when $Q_c = K_c$, the system is at equilibrium! So let's look at the given value of Q_c and compare it to the K_c that was provided for this reaction (EQ 4.4) at the given temperature of 90°C:

$Q_c = 0.010$ and $K_c = 2.39$, thus there are fewer product molecules (value of Q_c) than would be found if the system was at equilibrium (value of K_c).

What effect does this non-equilibrium condition have on this reversible reaction? The answer is provided by the *LeChatelier's Principle*, which is described in detail in your textbook. This principle states that: "when a stress is applied to a system at equilibrium, the system will shift to minimize the effect of the stress." So when Q_c is *less than* K_c, this creates a "stress" because there are not enough product molecules with respect to the equilibrium condition. Next, the system will shift to relieve that stress. However, what does that mean? Since the stress is "too few" product molecules—the system shifts to make more product molecules. How does this "shift" occur? The answer is that the rate of the forward reaction must be greater than the rate of the reverse reaction, $R_{forward} > R_{reverse}$, such that more product molecules will be made and more reactant molecules will be consumed. Will this process continue forever? No! It will stop when $Q_c = K_c$, which means that the system has reached equilibrium. Suppose that we find out that there are now ten molecules of each product (SO_2 and Cl_2) and ninety molecules of product (SO_2Cl_2). Do these amounts constitute an equilibrium condition? Let's find out:

$$Q_c = \{[10] * [10]\}/[90] = 100/90 = 1.11$$

Given that $K_c = 2.39$, $Q_c < K_c$, there are still not enough product molecules to constitute an equilibrium condition. What happens next? The answer is $R_{forward} > R_{reverse}$ continues as a non-equilibrium condition. Suppose we find out that there are now 60 molecules of SO_2Cl_2 and 10 molecules each of SO_2 and Cl_2. Is the system at equilibrium?

$$Q_c = \{[10] * [10]\}/[60] = 100/60 = 1.67$$

Thus, $Q_c < K_c$; however the value of Q_c is much closer to that of K_c. Finally, 65 molecules of SO_2Cl_2 and 12 molecules each of SO_2 and Cl_2. Is the system finally at equilibrium?

$$Q_c = \{[12] * [12]\}/[65] = 144/65 = 2.22$$

Actually, this value of Q_c is very close to the K_c for this reaction ($K_c = 2.39$). How could we get a more accurate value? The answer is that chemical reactions in the lab always involve at least millions and billions of molecules (even as high as $6.02 * 10^{23}$ molecules). So if we started with 1,049 molecules of SO_2Cl_2, then equilibrium would be $K_c = \{[49] * [49]\}/1000 = 2.40$ (very close to $K_c = 2.39$).

Overall—do you get the "concept" of chemical equilibrium? By "counting molecules" we hope you can understand the concept of *chemical equilibrium*. If you say "yes," then you can now work some mathematical problems with molarity (e.g., 0.10 M = [0.10]) or with partial pressures (e.g., $P_{gas} = 0.55$ atm). Note that your textbook clearly distinguishes between K_c when calculated from [molar], and K_p for partial pressures. That is, the relationship between these two equilibrium constants is based on PV = nRT, which is rearranged to P = (n/V)*RT or P = [M]*RT. Thus, the resulting relationship is $K_p = K_c(RT)^{\Delta n}$ where R = 0.0821 L*atm/(mol*K), T = temperature in Kelvin, and $\Delta n = n_{prod} - n_{reac}$. How can you calculate Δn? Answer: To find Δn, take the sum of coefficients of the product gases and *subtract* the sum for the reactant gases:

EQ 4.4: $SO_2(g) + 2\,H_2S(g) \longleftrightarrow 3\,S(s) + 2\,H_2O(g)$ $\quad$ $\mathbf{\Delta n = 2(g) - 3(g) = -1}$

For other mathematical variations—refer to your textbook.

Sec 4.8 Summary

The title of most textbook chapters on this topic is "chemical equilibrium." However, this title is deceptive because of several factors:

- A chemical reaction that has both forward and reverse reactions occurring: R → P & P → R is termed "reversible reaction" and not "chemical equilibrium";
- When most reversible reactions begin (t = 0), the system is at *non-equilibrium* because normally there is a huge excess of the reactants and no products; thus, $R_{forward} >> R_{backward}$
- If one side (reactant or product) has a much, much higher chemical potential energy than the other, then the reaction is irreversible (either an endo- or exo-thermic reaction).
- If the two sides (reactant and product) are roughly similar in their chemical potential energies, then the reaction is reversible:
  - If the value of K_c (the equilibrium constant) is much, much larger than one ($K_c >> 1$), then
    - The product(s) are much more stable than the reactant(s), which are reactive (unstable);
    - There are many, many more product molecules (ions or atoms) than there are reactant molecules.
  - If the value of K_c (the equilibrium constant) is much, much smaller than one ($K_c << 1$), then
    - The product(s) are much less stable than the reactant(s), which are reactive (unstable);
    - There are many, many more reactant molecules (ions or atoms) than there are product molecules.
  - If the value of K_c is moderately near one ($10^{-3} < K_c < 10^3$), then
    - The product(s) have similar stability to that of the reactant(s), if $K_c > 1$, then the products are moderately more stable; whereas, if $K_c < 1$, then the products are moderately less stable;
    - There are roughly the same numbers of reactant molecules (ions or atoms) as there are product molecules; however, $K_c > 1$ means there should be moderately more product molecules, while $K_c < 1$ means moderately fewer product molecules.

- If a stress, such as $Q_c \neq K_c$, then there are either . . .
  - Too many product molecules with respect to equilibrium when $Q_c > K_c$;
  - Too few product molecules with respect to equilibrium when $Q_c < K_c$;
  - This is one aspect of LeChatlier's principle that your textbook discusses in detail.
- Of course, the word "equilibrium" means "dynamic equilibrium," which means that $R_{forward} = R_{reverse}$; do you understand this concept? If "no," then
  - Think about a hillbilly feud where the McCoys are shooting at the Hatfields—at first one family and their kin are firing may more rounds at the other (non-equilibrium firefight); however, the other side will bring in more of their kin (and their guns) until the number of rounds fired back and forth are equal (equilibrium firefight).
- If you do not like this analogy, please invent your own . . . ____________ ☺

Overall, please remember that when you are trying to *understand* equilibrium and non-equilibrium—think of it as being a "*counting exercise.*" If this idea makes sense, then you have the concept of "equilibrium in chemical systems"!

Sec 4.9 Technical References

[1] Wettack, F. S. A photometric study of the N_2O_4—NO_2 equilibrium. A physical chemistry experiment. *J. Chem. Educ.*, **1972**, *49* (8), p. 556.

[2] Leenson, I. A. Approaching equilibrium in the N_2O_4—NO_2 system: A common mistake in textbooks. *J. Chem. Educ.*, **2000**, *77*(12), p. 1652.

[3] Reiffenstein, R. J., Hulbert, W. C., & Roth, S. H. Toxicology of hydrogen sulfide. *Annu. Rev. Pharrnaco Tl.oxicol.* **1992,** 109–134.

UNIT 5
Acid-Base Aqueous Systems

WHY STUDY THIS UNIT?

The goal of this unit is to help you develop a qualitative understanding of acid-base chemical systems. Specifically, this unit will cover how to . . .

- Understand different types of acids with respect to various degrees of acidity;
- Understand different types of bases in terms of their basicity;
- Visualize the pH scale in order to use it in meaningful ways;
- Visualize weak acid systems across their pH range to "see" buffer regions, and to double-check your calculations.

Your textbook usually devotes two chapters on topics involved in acid-base equilibria (e.g., Chapter 15 and 16).

Sec 5.1 Introduction

The "relative reactivity" of all acid species depends on the amount of hydronium ion they form in aqueous solutions. Hydronium ion is H_3O^+, and it is the strongest acid possible in water. A *strong acid* is fully ionized, and it has a higher chemical potential energy (larger circle) than that of the hydronium ion (smaller circle). This is because a strong acid (e.g., HCl (aq)), reacts with water to produce hydronium ion and chloride anion:

EQ 5.1: *Strong acid:* $HCl\,(aq) + H_2O\,(l) \rightarrow H_3O^+\,(aq) + Cl^-\,(aq)$

Higher Chem P.E.   *Lower Chem P.E.*

At equilibrium there would be many more hydronium ions than the number of HCl (aq) molecules.

As compared to hydronium ion, weak acids have much lower chemical potential energies, and hence are represented by smaller circles. Hydrofluoric acid, HF (aq), is a weak acid, which means that at equilibrium there would be many, many more HF molecules than there are hydronium ions.

EQ 5.2: *Weak acid:* $HF\,(aq) + H_2O\,(l) \leftrightarrow H_3O^+\,(aq) + F^-\,(aq)$

Higher Chem P.E.   *Lower Chem P.E.*

You may wonder why HCl (aq) is a strong acid and HF (aq) is a weak acid. The answer is that the covalent bond in H—F is much stronger than the one in the H—Cl bond. A stronger bond means that it is harder to break apart. H—F is a much smaller molecule than is H—Cl, thus the distance between the two nuclei in H—F is *much shorter* and the attraction of the two nuclei for the bonded electron pair in H—F is much greater than it is in H—Cl. So why is HCl (aq) called a "strong acid" when its covalent bond is much weaker than the HF bond? The reason is that a strong acid is actually an acid that is a *strong electrolyte*, and a strong electrolyte is 100% ionized. Thus, 100% of HCl molecules are ionized to form H^+ (aq) and Cl^- (aq). Conversely, HF (aq) is a weak acid because it has a strong covalent bond that keeps most of HF in molecular form. Thus, only a small fraction of HF (aq) molecules are broken into ions. Since HF only gives a dim glow in the conductivity apparatus, it is a *weak electrolyte*: HF (aq) $\leftrightarrow$ H^+ (aq) + F^- (aq). That is, hydrofluoric acid is dissolved mostly as molecules and only partially as ions.

Bases are chemical species that generate hydroxide ion, OH^-, in aqueous solutions. Strong bases are ionic compounds that dissolve in water and produce hydroxide ion. Thus, soluble metal hydroxides are strong bases because they ionize completely (~100%) to yield soluble metal ions and hydroxide ions. When you look at a table of solubility rules for ionic compounds, most metal hydroxides are insoluble in water; however, Group 1 MOHs and some Group 2 MOHs are soluble. Specifically, LiOH, NaOH, KOH, RbOH, are soluble in water and are, hence, strong bases. In addition, $Ca(OH)_2$, $Ba(OH)_2$, and $Sr(OH)_2$ are all soluble compounds, hence strong bases. These strong bases are completely ionized (~100%) in water and are thus strong electrolytes. For example,

EQ 5.3: *Strong base* $Ca(OH)_2\,(s) \xrightarrow{H_2O\,(l)} Ca^{2+}\,(aq) + 2\,OH^-\,(aq)$

Conversely, weak bases are only partially ionized, and are thus weak electrolytes. The most common weak base is aqueous ammonia, NH_3 (aq). When you write this reaction, you must include water as a reactant that reacts with ammonia to form ammonium hydroxide:

EQ 5.4: *Weak base* $NH_3\,(aq) + HOH\,(l) \leftrightarrow NH_4^+(aq) + OH^-(aq)$

$NH_4OH\,(aq) \leftrightarrow NH_4^+(aq) + OH^-(aq)$

Lower Chem P.E. *Higher Chem P.E.*

Aqueous ammonia has two distinct formulas—NH_3 (aq) & NH_4OH (aq). This is probably true because this is a reversible reaction. You should expect the equilibrium constant to be less than one because of the higher chemical potential energy of the hydroxide ion product with respect to the lower chem PE's of the reactants. Most of the other weak bases are derivatives of ammonia—where one to three H's are replaced—for example, CH_3NH_2 (methyl amine) replaces one H with a CH_3 group, which does not affect the basicity of these compounds (i.e., it is still a weak base). Likewise, dimethyl amine, $(CH_3)_2NH$, and trimethyl amine, $(CH_3)_3N$, are weak bases in water. All weak bases are only partially ionized—weak electrolytes in water. At equilibrium in EQ 5.4, you should expect to see many aqueous ammonia molecules but very few OH– ions.

You are probably familiar with the pH scale from your previous coursework, especially in biology. Thus, you know that pH = 7 is neutral and water, HOH, is neutral, while a pH value that is greater than 7 is a basic solution, and less than 7 is an acidic solution. In Sec 5.4 we will explore how you can visualize the pH scale. In a preview of this section, the pH scale is a logarithmic scale—so if the pH is lowered from 6 to 5, the solution is not twice as acidic but rather it is actually ten times more acidic. Likewise, when pH is increased from 8 to 9, the solution is ten times more basic.

An acid-base neutralization reaction is a reaction between an acid, HX (aq), and a base, MOH (aq), which yields a salt, MX, plus water. In this reaction, the acid and base reactants are "more reactive" (less stable), and the products (salt and water) as "more stable" (less reactive). For example,

EQ 5.5: HX (aq) + MOH (aq) ➔ HOH (l) + MX (aq)

HX (aq)	MOH (aq)	HOH (l)	MX (aq)
Higher Chem P.E.	*Higher Chem P.E.*	*Lower Chem P.E.*	*Lower Chem P.E.*

You should recognize this reaction as being an irreversible reaction (R ➔ P not R ⟷ P). How would you draw circles to represent the chemical potential energies of these chemical species (HX, MOH, HOH, and MX)? If you are thinking larger circles for HX (aq) and MOH (aq) and smaller ones for HOH (l) and MX (aq), then you are correct! Do you also see that in terms of the law of conservation of energy, that something is missing on the product side? What is it? The answer is "heat" is always produced by one of these neutralization reactions. In other words, all neutralization reactions are exothermic!

Sec 5.2 Acids and Degree of Acidity

Are you familiar with some of the features of aqueous acids? First, an acid donates a proton in a chemical reaction, and a proton is shown as H^+. Also, the chemical formula for most acids have a proton as the cation: HX or HXO_n. Most textbooks often refer to acidic conditions as containing H^+ (aq) ion. How does this representation relate to acids that are dissolved in water? The answer is described in terms of *net ionic equation*, which you studied in your previous chemistry course. For example,

EQ 5.6: *Strong acid ionization:* HX (aq) + H_2O (l) ➔ H_3O^+ (aq) + X^- (aq)

Subtract H_2O from both sides: HX (aq) ➔ H^+ (aq) + X^- (aq)

So in terms of chemistry notations, hydronium ion and hydrogen ion are equivalent structures. Frequently, chemists use hydrogen ion, H^+ (aq), as an abbreviation for hydronium ion, H_3O^+ (aq), just like ATP is an abbreviation for Adenosine TriPhosphate. However, these two notations for acidity do not have the same chemical potential energies:

EQ 5.7: H^+ (aq) + H_2O (l) ➔ H_3O^+ (aq)

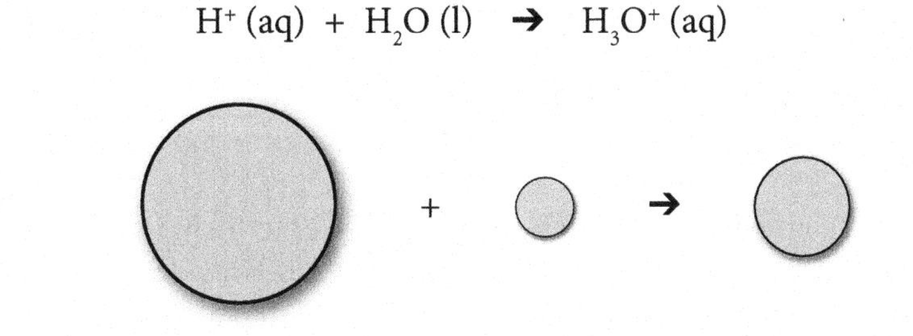

A hydrogen ion is often referred to as a *proton* because when an electron is removed from a hydrogen atom, there are no electrons left in the hydrogen ion. Thus, hydrogen ion is just the nucleus of hydrogen, which in the most common isotope of hydrogen (hydrogen-1, 99%) is just a proton

with no neutrons. Looking at the above diagram—if a proton were to suddenly appear in water, it would immediately react with one of the nonbonded electron pairs on the oxygen of water. $H^+ + :OH_2$ (l) → $H:OH_2^+$ (aq) where $:OH_2$ (l) is written to show the electron pair that reacts with the proton, and $H:OH_2^+$ (aq) is just another way to write H_3O^+ (aq). Therefore, in terms of chemical potential energies hydrogen ion H^+ (aq) does not exist because the addition of a proton, H^+, to water would react spontaneously. Why? No bonds are broken in this reaction, and a covalent bond is formed between hydrogen and the electron pair of oxygen. That is, there is no energy cost to this reaction—only energy released from the formation of the strong covalent bond:

EQ 5.8: $$H^+ \text{ (aq)} + H_2O \text{ (l)} \rightarrow H_3O^+ \text{ (aq)} + \text{heat energy}$$

The above diagram should make sense to you because it obeys the *law of conservation of energy*. Also, as described in Unit 2 (chemical kinetics) when no bonds need to be broken—there is no activation energy, in other words: E_{act} = zero for EQ 5.8.

So, how do we represent chemical potential energies of the various types of acids? Let's compare the chemical PE's of strong acids (~100% ionized), weak acids (< 5% ionized; > 95% molecular), and water:

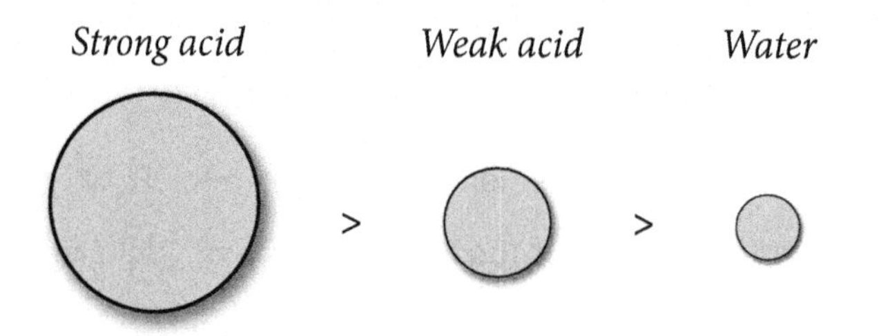

In water, all three of these chemical species produce hydronium ion, H_3O^+ (aq). However, the extent to which this reaction occurs varies in a systematic manner:

EQ 5.9: *Strong acid:*	HX (aq)	+ H_2O (l)	→ H_3O^+ (aq)	+ X^- (aq)
Before it dissolves:	100%	----	0%	0%
After it dissolves:	~0%	----	100%	100%

As shown in EQ 5.9, a strong acid completely ionizes in water to form its ionized products, hydronium ion and that acid's anion (e.g., for HCl (aq), the anion is Cl^-). How do chemists know that ionization is complete—100% or very nearly 100%? The answer is obtained as follows:

- Use a conductivity apparatus, which consists of a light bulb but which has an incomplete electrical circuit that is then submerged into an aqueous solution;

- Any electrolyte completes the circuit and produces a glow in the bulb, which means that some (or all) of the molecules are broken down into ions;
- When you dissolve acid molecules in water and you see a bright glow, then that acid is a strong acid, which is completely ionized, 100% (its name is derived from "acid that is a strong electrolyte").

Conversely, if you dissolve "weak acid" molecules in water, you see only a dim glow, which indicates that the acid dissolves mostly in molecular form rather than in ionized form.

EQ 5.10: *Weak acid:*	HA (aq)	+ H_2O (l)	⟷	H_3O^+ (aq)	+ A^- (aq)
Before it dissolves:	100%	----		0%	0%
After it dissolves:	>95%	----		< 5%	< 5%

The double-headed arrow (⟷) in EQ 5.10 means that the reaction occurs only to a limited extent: weak acids are "weak electrolytes" that are only *partially ionized.* You will learn why "less than 5% ionized" is the proper quantity when this topic is described in detail in your textbook (a chapter on acid/base equilibria).

In pure water, ionization to produce hydronium ion does occur; however, for every positive H_3O^+ formed, there is a hydroxide ion, OH^- formed. As discussed in your textbook, pure water is neutral because it forms equal amounts of acidic hydronium ion and basic hydroxide ion.

EQ 5.11: *water:*	HOH (l)	+ HOH (l)	⟷	H_3O^+ (aq)	+ OH^- (aq)
Before it ionizes:	100%	----		0%	0%
After it ionizes:	~100%	----		~0% *	~ 0% *

In pure water at room temperature, 25°C, only one out of every ten million (1/(10^7) water molecules is ionized:

$$[H_2O] = (1 \text{ mol}/18.0 \text{ g}) * (1000 \text{ mL/L}) = 55.5 \text{ mol/L}$$

$$[H_3O^+] = 10^{-7} \text{ M} \quad \text{and} \quad [OH^-] = 10^{-7} \text{ M}$$

How small is the number 10^{-7}? It can also be written as 1/(10^7) or 1/10,000,000. Thus, only one molecule of water (in pure water) in ten-million molecules breaks up into ions as shown in EQ 5.11. In the real world, suppose you and ten million other people purchased lottery tickets; what would be the odds of you winning a megabucks lottery? Answer one in ten million, in other words 9,999,999 people would be holding worthless tickets (assuming that only one monetary prize is given). So in pure water the vast number of water molecules are molecular species, which are not ionized.

Sec 5.3 Bases and Their Degree of Ionization

All strong bases are soluble metal hydroxides. Specifically, these bases include Group I hydroxides (e.g., LiOH, NaOH, KOH, etc.) and "CBS" meaning $Ca(OH)_2$, $Ba(OH)_2$, and $Sr(OH)_2$. All strong bases are 100% ionized. Thus, if you have 0.100 M KOH, the $[OH^-]$ = 0.100 M. What is the pH of this solution? First, pOH = $-\log[OH^-]$ and so pOH = $-\log[0.100]$, and pOH = 1.0. Next, as your textbook indicates: 14 = pH + pOH, so pH = 14 – pOH, and pH = 14 – 1.0 = 13.0.

All common weak bases are related to ammonia, NH_3 (aq). You need to remember to include water in the equilibrium expression for a weak base:

EQ 5.12: $NH_3\ (aq) + H_2O\ (l) \leftrightarrow NH_4OH\ (aq) \leftrightarrow NH_4^+(aq) + OH^-(aq)$

$K_b = 1.8 * 10^{-5}$

Typically, the concentration of OH^- must be calculated using the K_b value:

$$K_b = \frac{[NH_4^+]_{equil} * [OH^-]_{equil}}{[NH_4OH]_{equil}} = \frac{[NH_4^+]_{equil} * [OH^-]_{equil}}{[NH_3\ (aq)]_{equil}}$$

So, if $[NH_3\ (aq)] = 0.0100$ M—use the ICE method—described in your textbook to give . . .

$$1.8 * 10^{-5} = \frac{x * x}{[0.0100 - x]}$$

$$x^2 = [0.0100] * 1.8 * 10^{-5} = 1.8 * 10^{-7}$$

$$x = (1.8 * 10^{-7})^{½} = (1.8 * 10^{-7})^{0.5} = 4.2 * 10^{-4}$$

$$x = [OH^-]_{equil} = 4.2 * 10^{-4}\ M$$

$$pOH = -\log[4.2 * 10^{-4}] = 3.4, \text{ thus } pH = 14.0 - pOH = 14.0 - 3.4 = 10.6$$

For 0.010 M NH_3 (aq), pH = 10.6, which is more basic than a 0.010 M KOH, which has pH = 12.0. The pH of the strong base is twenty-five times ("log 1.4" = 25) more basic than the pH of ammonia, a weak base. Next, we need to discuss the pH scale, which is valid for all aqueous solutions at a temperature of 25°C (~room temperature).

Sec 5.4 Visualizing the pH Scale

As mentioned in the introduction, Sec 5.1, most students are familiar with the term "pH" as it relates to three regions: pH < 7 → acidic; pH = 7 → neutral; pH > 7 → basic. However, do you understand this logarithmic scale? First, the extreme values of the pH scale represent large concentrations of acid (e.g., pH = 1 or 2) or base (e.g., pH = 13 or 14). Next, look at the mathematical equation for pH:

$$pH = -\log[H^+]$$

This equation can be rearranged to give $10^{-pH} = [H^+]$. This version of the logarithmic equation shows what kind of number pH really is . . . pH is the exponent that expresses the $[H^+]$ concentration. Why is pH equal to the negative log? The answer is just that whoever invented the concept of pH got really tired of writing all the negative numbers on the scale: 0 –1 –2 –3 –4 –5 –6 –7 –8, and so on. However, looking at the negative scale can help you better understand pH as a logarithmic scale (e.g., the number $10^{-7} = 0.0000001$, while $10^{-4} = 0.0001$, and $10^{-1} = 0.1$). Thus, it is easier to see that *smaller pH numbers* represent *larger concentrations* of $[H^+]$. Next, let's compare pH = 4.0 with pH = 1.0 as follows: pH = 4.0, so $10^{-pH} = 0.00010$ M H^+, and when pH = 1.0, then: $10^{-pH} = 0.10$ M H^+. Therefore, when we go from pH = 4.0 to pH = 1.0, there is a thousand-fold increase in acidity. Can you appreciate this 1000X increase? What if you are flipping burgers at MacBurgers and making the minimum wage of $8.25 per hour. You ask your boss, who does not understand logarithms, for a pay raise of just +3 on the log scale. What is your new hourly wage? New wage = $8.25 * 10^3 = $8.25 * 1000 = $8,250 per hour! If you get this new contract, then you would not have to work at MacBurgers for more than a couple of weeks (pre-tax earnings = $660,000) or maybe a month (earnings = $1,320,000). So let's stop dreaming and get back to chemistry! Any change of 3.0 that occurs from a larger pH to a smaller pH is an increase in acidity of 1,000 times. Likewise, going up the pH scale from 9 to 12 is an increase in basicity of 1,000 times.

An alternative way to visualize the pH scale is via a colorimetric pH scale as shown in Figure 5.1 below. First, notice the three regions of colors: pH 1 to 4 are "warm colors," pH 5 to 8 are "shades of green," and pH 9 to 12 are "purplish." Also, notice that with a keen eye, you can distinguish the pH to within 1.0 unit. This type of colorimetric pH scale can be obtained by adding several drops of "Universal Indicator" (UI), which shows that changes in color are matched to the changes in pH. UI is actually a mixture of a bunch of indicators to achieve the three regions of colors. For example, you may have previously used phenolphthalein for acid-base titrations. It is colorless in the acidic range and purple in the more basic ranges, which begins at about pH = 9.

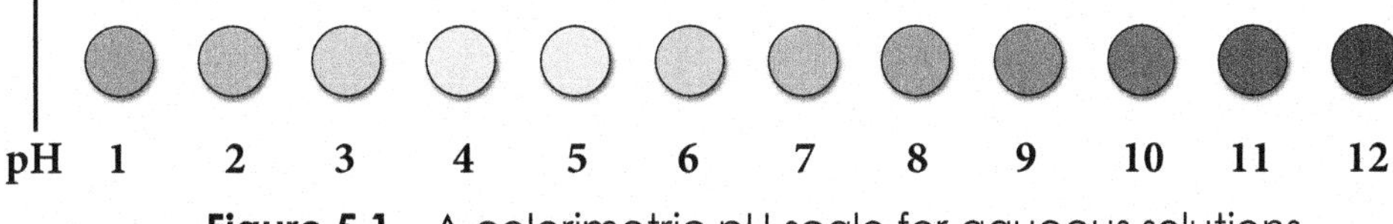

Figure 5.1 A colorimetric pH scale for aqueous solutions

When you are doing mathematical calculations with pH, you should always estimate what the pH will be—even before you punch your calculator. For example, given that hydrogen ion concentration, $[H^+] = 2.5 * 10^{-4}$ M, then you know that the pH is the exponent but ignore the negative sign. Therefore, pH ~ 4 and furthermore, since $10.0 * 10^{-4}$ M is equal to $1.00 * 10^{-3}$ M, this means that the pH is between 3.0 and 4.0. Now, calculate the pH with your calculator: pH = – log $[H^+] = 2.5 * 10^{-4}$, enter the number, $2.5 * 10^{-4}$, then press the log key to get . . . –3.602, ignore the negative sign and round off to get . . . pH = 3.6. This value is within the range of possible pH's 3.0 to 4.0, thus it is a reasonable answer. When you calculate pH values, please do these estimations so you can understand your answer and find any errors that were made.

Calculating the pH of a *strong acid* is very easy because it is 100% ionized by definition. Thus, given $[HNO_3] = 5.0 * 10^{-3}$ M, then $[H^+] = 5.0 * 10^{-3}$ M, and pH = –log [H+] = –log $[5.0 * 10^{-3}]$ = 2.3 (note that the answer is between 2.0 and 3.0). To calculate the pH of a weak acid when given the concentration of the weak acid—you need to first use the ICE method (I = initial concentration, C = change, and E = Equilibrium) to calculate the $[H^+]$. Your textbook will have several examples of how to do this type of problem. From $[H^+]_{equil}$ you have the information you need to calculate the pH for that weak acid. For example, given $[HNO_2]_{init} = 5.0 * 10^{-3}$ M (same as for the strong acid, HNO_3) and its K_a = $4.5 * 10^{-4}$, then ICE calculation gives an answer of $[H^+] = 1.5 * 10^{-3}$ M, and then pH = 2.8. So for the same concentration of a weak acid, which is only partially ionized to form $[H^+]$, there are less $[H^+]_{equil}$ and hence the pH is less acidic. What if another weak acid has a smaller K_a than the one for HNO_2?

For HClO (aq), which is hypochlorous acid, given $[HClO]_{init} = 5.0 * 10^{-3}$ M (same as above) and $K_a = 3.0 * 10^{-8}$ (much smaller value), then $[H^+]_{equil} = 1.2 * 10^{-5}$ M, and the corresponding pH = 4.9. So let's compare these two weak acids with a strong acid when they have *equal initial concentrations*:

- Strong acid, HNO_3, $[H^+]_{equil} = 5.0 * 10^{-3}$ M and pH = 2.3 (then H^+ = 400X)
- Weak acid, HNO_2, $[H^+]_{equil} = 1.5 * 10^{-3}$ M and pH = 2.8 (then H^+ = 126X)
- Weaker acid, HClO, $[H^+]_{equil} = 1.2 * 10^{-5}$ M and pH = 4.9 (if H^+ = 1X)

First, for every H^+ ion found in HClO, HNO_2 would have 126 times more H^+ ions, and HNO_3 400 times more H^+. How can you conceptualize an equilibrium constant that is very, very small? For example, what is the ratio of H^+ to HNO_2 molecules in nitrous acid, HNO_2 (aq), which has $K_a = 4.5 * 10^{-4}$? Let's round this number to $4 * 10^{-4}$ to illustrate how many particles are present at equilibrium:

$$K_a = \frac{[H^+]_{equil} * [NO_2^-]_{equil}}{[HNO_2]_{equil}} = \frac{[H^+]^2_{equil}}{[HNO_2]_{equil}}$$

Thus, the ratio of H^+ ions to HNO_2 molecules at equilibrium is equivalent to <u>two</u> H^+ ions per 10,000 HNO_2 molecules.

$$K_a = \frac{[2]^2}{[10{,}000]} = 4 * 10^{-4}$$

To really understand this ratio of $[H^+]$ to $[HNO_2]$ at equilibrium, you need to relate this ratio to their relative *chemical potential energies*. That is, H^+ ions are very unstable and hence should have a very large circle, while HNO_2 molecules are relatively stable—thousands of tiny circles. For a much weaker acid, HClO (aq), what would be the $[H^+]$ to [HClO] ratio? Given its $K_a = 3.0 * 10^{-8}$, how many particles are present at equilibrium:

$$K_a = \frac{[H^+]_{equil} * [ClO^-]_{equil}}{[HClO]_{equil}} = \frac{[H^+]^2_{equil}}{[HClO]_{equil}}$$

Thus, the ratio of H^+ ions to HClO molecules at equilibrium is equivalent to <u>three</u> H^+ ions per 30 million HClO molecules.

$$K_a = \frac{[3]^2}{[30{,}000{,}000]} = 3 * 10^{-8}$$

This value of K_a means the ratio of H^+ to HClO molecules is even much smaller than the one for the previous example, HNO_2. Why? The H^+ ion is equally reactive (unstable) in both acidic solutions; and the molecule of HClO would be even more stable. In other words, if it were possible to search for H^+ ions at the molecular level, the odds of finding one among the acid molecules would be one out of 10 million (3 out of 30 million).

Sec 5.5 Monoprotic Acid Distribution Systems across pH

Monoprotic acid are strong acids that have very simple distribution systems across all the pH's found in aqueous solutions:

Strong acid	HX (aq)	➔	H^+ (aq) +	X^- (aq)
Start:	100%		0%	0%
Equilibrium:	0%		100%	100%

Thus, if you know the concentration of the molecular HX, then the concentration of $[H^+]_{equil}$ = $[HX]_{initial}$. For example, $[HNO_3]_{initial}$ = 0.10 M, then $[H^+]_{equil}$ = 0.10 M. Also, $[X^-]_{equil} = [HX]_{initial}$, thus $[NO_3^-]_{equil} = [H^+]_{equil}$ = 0.10 M

Conversely, the concentration of a weak acid (e.g., $[HF]_{equil}$, and its anion $[F^-]_{equil}$) are dependent on the pH of the system, which is, in turn, dependent on the value of K_a for a particular weak acid.

For example, the equation for K_a of a weak acid can be transformed into an equation that includes both the pH, which can vary, and the pK_a, which is fixed for a given weak acid. This relationship is called the *Henderson-Hasselbalch equation*:

$$pH = pK_a + \log [A-]/[HA]$$

We need to solve this equation for the logarithm of this ratio: $[F^-]_{equil}$ to $[HF]_{equil}$

$$\log [A^-]/[HA] = pH - pK_a$$

However, does this make any sense to you? To put it another way, can you see what happens to this ratio as pH is systematically varied? If you are drawing a "blank" right now, don't feel alone!

Fortunately, there is a visual way to "see" how this ratio varies with changes in the pH of the weak acid system. For example, let's look at benzoic acid, $HO_2C(C_6H_5)$, which we will abbreviate as HOBz, where $OBz^- = O_2C(C_6H_5)^-$ anion (see Figure 5.2). Next, given that the equilibrium constant for HOBz is . . . $K_a = 6.3 * 10^{-5}$, we can calculate pK_a as follows:

$$pK_a = -\log (6.3 * 10^{-5}) = 4.2$$

Why is pKa the most important feature in visualizing a weak acid system? First, let's look at...

- If pH = pK_a, then: $[HOBz]_{equil} = [OBz^-]_{equil}$, thus, this is the "pivotal point";
- If pH < pK_a, then $[HOBz]_{equil} > [OBz^-]_{equil}$, thus, HOBz is the *dominant species;*
- If pH > pK_a, then $[HOBz]_{equil} < [OBz^-]_{equil}$, thus, OBz^- is the *dominant species.*

Now, look at Figure 5.2 for the benzoic acid system. First, trace the <u>protonated species</u> [HOBz], from pH = 0.0 to pH = 2, where [HOBz] is ~ 100% then it drops to 90% at pH = 3.2 (1.0 pH unit below pK_a), where the ratio of [HOBz] :$[OBz^-]$ is 10:1. Next, the HOBz concentration declines to 50% at pH = pK_a = 4.2. Continue to follow the [HOBz] curve beyond pK_a to higher pH's. At pH = 5.2 (pK_a + 1.0), the [HOBz] :$[OBz^-]$ ratio has dropped to 1:10 (only 10% HOBz). Continuing on to pH = 6.2 (pK_a + 2.0), the [HOBz] :$[OBz^-]$ ratio is 1:100 ratio and [HOBz] is only 1.0%. At higher pH values than 6.2, the concentration of HOBz approaches zero.

Now using Figure 5.2, we can follow the concentration of the *deprotonated species*, $[OBz^-]$ from the most basic pH, pH = 14, down to where pH = pKa. Note that from pH 14 to pH 6.2, the concentration of $[OBz^-]$ is essentially 100% (and [HOBz] is ~zero%). At pH = 6.2 (pK_a+2), the concentration of $[OBz^-]$ is ~99% (a 100:1 ratio of $[OBz^-]$:[HOBz]). Then at pH = 5.2 (pK_a+1), the concentration of $[OBz^-]$ is ~90% (a 10:1 ratio of $[OBz^-]$:[HOBz]). Finally, at pH = pK_a, $[OBz^-]$ is 50% (a 1:1 ratio of $[OBz^-]$:[HOBz]).

What does it take to make a "*good buffer*" from a weak acid and its anion? The answer is that there needs to be roughly similar amounts of both of these chemical species: The best buffer is where pH = pK_a, then $[HOBz]_{equil}$ = $[OBz^-]_{equil}$. This is a good buffer solution because it can react with a strong base, OH^-, by converting it to a weak base, OBz^-:

EQ 5.13:

HOBz +	OH^- (aq)	➔	HOH (l) + OBz^-
Weak acid	*Strong base*	➔	*Weak base*

The pH does rise slightly but not near as much as when a strong base is added to a non-buffered solution. That is, the addition of 100% OH^- would greatly increase the pH of the solution, while the weak base (~5% OH^-) produces only a slight increase in pH. This also accounts for the buffering effect of a strong acid: when a strong acid, H^+ (aq), is added to this benzoate buffer solution:

EQ 5.14:

OBz^- +	H^+ (aq)	➔	HOBz
Anion	*Strong acid*	➔	*Weak acid*

Overall, for a monoprotic weak acid, the region of a HA/A^- system that is a good *buffering solution* is always found within 1.0 pH unit of the pK_a. Again, this is because the buffer contains both acid, HA, and base, A^-, components that can react with either a strong base or a strong acid to "resist changes in pH."

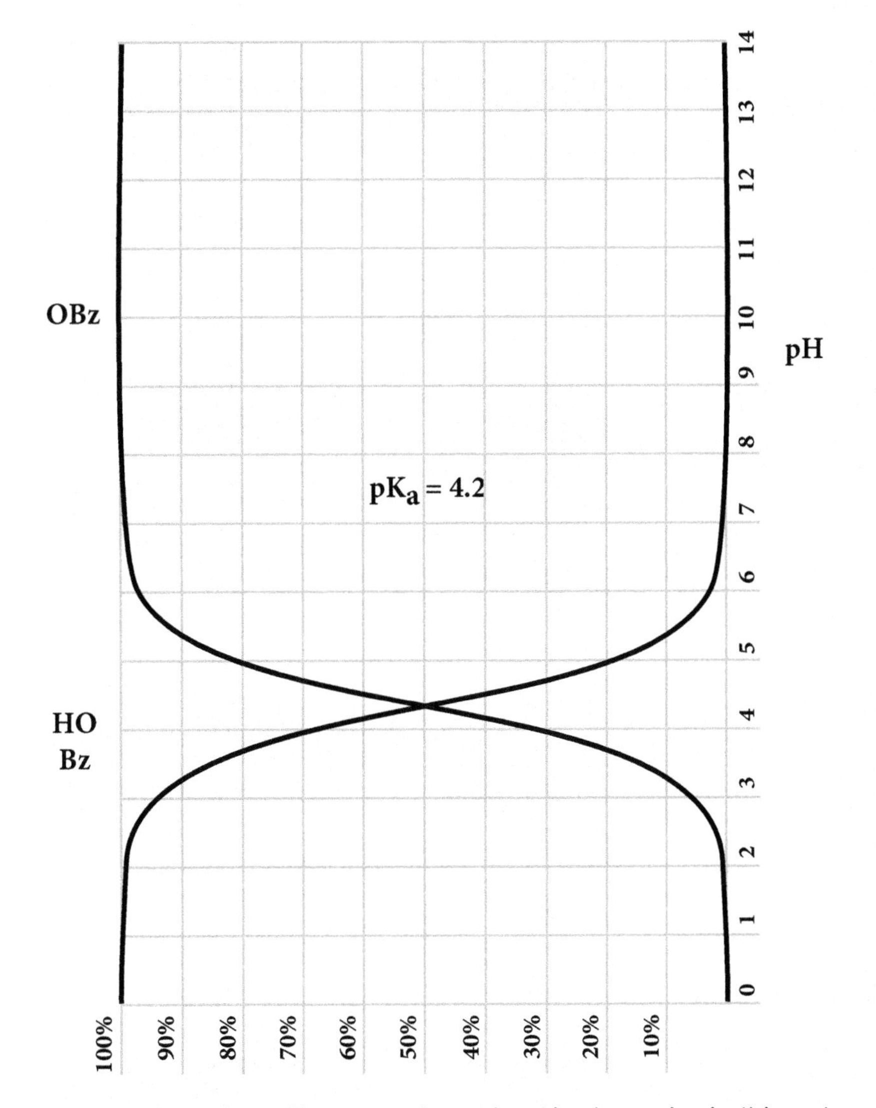

Figure 5.2 *Benzoic acid system*. There are <u>two</u> chemical species in this system:

- HOBz, the dominate species at pH < pK_a (pK_a = 4.2)
- OBz– at pH's > pK_a.

Sec 5.6 Diprotic Acid Distribution Systems across pH

As you may recall from reading the acid/base equilibrium chapter of your textbook: In an acid-base neutralization reaction, a *diprotic acid*, H_2A (generic formula), can contribute two protons per mole of acid. Right away, you may think of sulfuric acid, H_2SO_4 (aq) {the acid found in car batteries}, as being an important diprotic acid. Yes, it certainly is one; however, it is a little more complex because the first proton to dissociate from it is 100% ionized (i.e., strong acid). Conversely, the second proton only partially dissociates:

EQ 5.15A: $H_2SO_4(aq) \rightarrow H^+(aq) + HSO_4^-(aq)$ K_1 = very large (100% ionized)

EQ 5.15B: $HSO_4^-(aq) \rightarrow H^+(aq) + SO_4^{2-}(aq)$ $K_2 = 1.2 * 10^{-2}$

So let's not use sulfuric acid for our diprotic acid.

There is another diprotic acid that has many real-world examples: the *carbonate system*: CO_2 (g), CO_2 (aq), H_2CO_3 (aq), HCO_3^- (aq), CO_3^{2-} (aq). This system is vitally linked to the chemistry of the world's oceans. That is, CO_2 (g) dissolves in ocean water to produce CO_2 (aq) and to react with water to produce carbonic acid:

EQ 5.16: $$CO_2(g) \xleftrightarrow{H_2O} CO_2(aq) + H_2O(l) \leftrightarrow H_2CO_3(aq)$$

Excess CO_2 (g) in the atmosphere is a major factor in the acidification of our oceans [1,2,3]. Is the carbonate system dependent on pH? Yes!

As shown in Figure 5.3, you can determine the dominant species in the carbonate system as a function of pH. However, where are the "pivotal points" of this system? The answer is that the two pKa's of this system are these points:

EQ 5.17A; $H_2CO_3(aq) \leftrightarrow HCO_3^-(aq)$ $K_1 = 4.3 * 10^{-7}$ $pK_1 = -\log(4.3 * 10^{-7}) = 6.4$

EQ 5.17B: $HCO_3^-(aq) \leftrightarrow CO_3^{2-}(aq)$ $K_2 = 5.6 * 10^{-11}$ $pK_2 = -\log(5.6 * 10^{-11}) = 10.2$

The dominant species are as follows (Figure 5.3):

pH:	0.0 to 6.4	H_2CO_3 (aq)
pH:	6.4 to 10.3	HCO_3^- (aq)
pH:	10.3 to 14.0	CO_3^{2-} (aq)

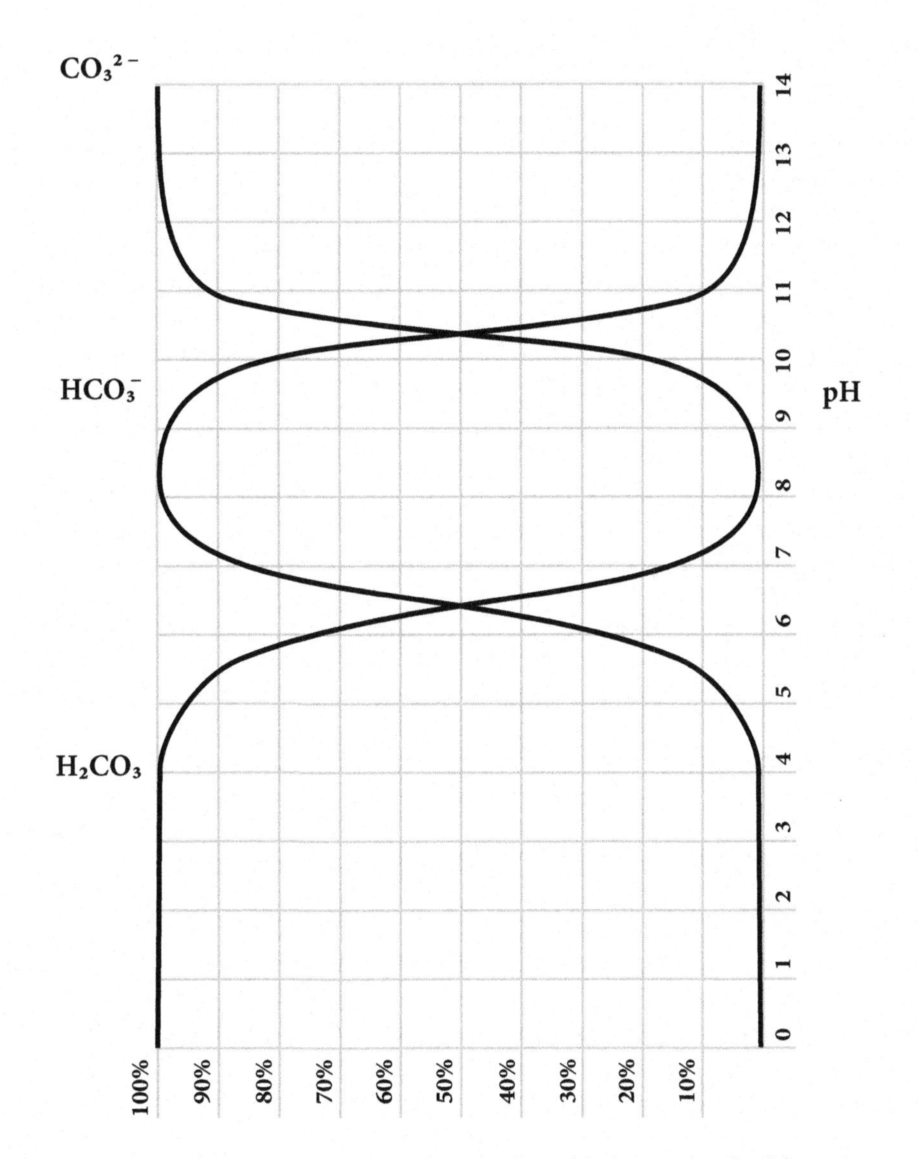

Figure 5.3 *Carbonate system.* There are three chemical species in this system:

- H_2CO_3, dominates at pH's < pK_1 (where $pK_1 = 6.4$),
- HCO_3^- dominates at pH's between pK_1, and pK_2 , and
- CO_3^{2-} at pH's > pK_2 (where $pK_2 = 10.3$).

Thus, you can measure the pH of a carbonate solution and then use Figure 5.3 to determine the dominant species at that pH. Suppose you are brewing homemade beer and you need a mash pH = 5.3 for the carbonate system[4]. If the pH is currently pH = 6.3, then you need to add an acid until the pH is adjusted to pH = 5.3. Brewers often use lactic acid, a weak acid, for this purpose. Referring to Figure 5.3, we can determine the carbonate species at this mash pH. Thus, at pH = 5.3, H_2CO_3 is the dominant species and it constitutes ~90%, while HCO_3^- (aq) is ~10%. What about the remaining species, CO_3^{2-}? At the mash pH of 5.3, $[CO_3^{2-}]$ is far less than 1%.

Notice that where the pH is alkaline (pH > 7), that CO_3^{2-} is the dominant species. What happens to $[CO_3^{2-}]$ when an acid is added to an alkaline solution?

- The $[CO_3^{2-}]$ concentration decreases by 10X for every drop of 1.0 pH unit from a given alkaline concentration of $[CO_3^{2-}]$.

Can you use Figure 5.3 when you are working mathematical calculations? Yes! One possibility is to use this diagram to help estimate your calculated answer. If you are given the pH, then you can determine which carbonate is the dominant species: H_2CO_3, HCO_3^-, or CO_3^{2-}. For example, given a pH = 9.0, what are the concentrations of these carbonate species? Looking at Figure 5.3, at pH = 9.0, HCO_3^- is the dominant species at (95%), CO_3^{2-} is ~5%, and H_2CO_3 is <1%. Next, if the total concentration of all carbonate species = 0.100 M, then $[HCO_3^-]_{equil} = 0.95 * 0.100\ M = 0.095\ M$, $[CO_3^{2-}]_{equil} = 0.05 * 0.100\ M = 0.005\ M$.

What do you do if you do not have a diagram of the pH distribution system for a particular weak acid? Let's say that you need to solve a problem with the tartaric acid, $H_2C_4H_4O_6$, (an acid found in grapes), and you are given the following information about this weak acid system:

EQ 5.18A: $H_2C_4H_4O_6\,(aq) \leftrightarrow H^+\,(aq) + HC_4H_4O_6^-\,(aq) \qquad K_1 = 1.0 * 10^{-3} \rightarrow pK_1 = 3.0$

EQ 5.18B: $HC_4H_4O_6^-\,(aq) \leftrightarrow H^+\,(aq) + C_4H_4O_6^{2-}\,(aq) \qquad K_2 = 4.6 * 10^{-5} \rightarrow pK_2 = 4.3$

Do you recognize this acid as a diprotic acid? What are the three dominant species with respect to pH ranges?

- From pH = 0.0 to 3.0, it is $H_2C_4H_4O_6$ (aq) the fully protonated species
- From pH = 3.0 to 4.3, it is $HC_4H_4O_6^-$(aq) the "protonated anion" species
- From pH = 4.3 to 14.0, it is $C_4H_4O_6^{2-}$ (aq) the deprotonated species

Given that the pH = 6.3 and total concentration of tartarate species = 0.080 M: What are the approximate concentrations of each species? First, $C_4H_4O_6^{2-}$ is the dominant species, and since pH = pK_2 + 2.0, its concentration can be estimated as follows:

- pH = pK_2 and 1:1 ratio, 50%, so $[C_4H_4O_6^{2-}]_{equil} = 0.50 * 0.080\ M = 0.040\ M$, then
- pH = pK_2 + 1.0 and 10:1 ratio of $[C_4H_4O_6^{2-}]:[HC_4H_4O_6^-] = 0.90 * 0.080\ M = 0.072\ M$,
- pH = pK_2 + 2.0 and 100:1 ratio of $[C_4H_4O_6^{2-}]:[HC_4H_4O_6^-] = 0.99 * 0.080\ M = 0.079\ M$

Estimated answer: $[C_4H_4O_6^{2-}] = 0.079$ M, and $[HC_4H_4O_6^-] = 0.001$ M (i.e., 1%). If you can make a habit out of doing these "estimations," then you can "double check" your calculated answers and you can begin to really understand diprotic acid systems!

Understanding the carbonate-bicarbonate buffering system of blood is essential in order to gain a biomedical understanding of human physiology. This system is linked to respiration where CO_2 is transported as part of this buffering system. When you are exercising, the lactic acid/lactate system, which includes the appropriate enzymes, regulates pH[5]. In human physiology, the *acid-base homeostasis* in the blood is vitally dependent on its carbonate and phosphate buffering systems.

Sec 5.7 Summary

Overall, if you can use these concepts to visualize a weak acid system across the pH scale, then you understand these equilibrium systems. You can use this system to solve a K_a problem in a meaningful manner. First, you should view the pH scale as a logarithmic scale where neutral (pH = 7) means [H_3O^+ or H^+] = [OH^-]. As the pH moves away from it, the solution either becomes more acidic (if pH < 7) or basic (if pH > 7). Next, you should convert K_a to pK_a to understand the ratio of weak acid (HA) to its anion (A^-). That is, at pH = pK_a, $[HA]_{equil} = [A^-]_{equil}$. If the pH is equal to pK_a – 1.0, then $[HA]_{equil} > [A-]_{equil}$, and the [HA] to [$A^-$] ratio is equal to 10:1. Likewise at pH = pKa + 1, then $[A^-]_{equil} > [HA]_{equil}$ and [A^-] to [HA] is equal to 10:1. As the pH drops by each 1.0 unit away from the pK_a, there is a 10X increase in [HA]. Likewise, as the pH increases by 1.0 unit, [HA] drops by 10X and the [A^-] increases by 10X.

Understanding *buffering systems* is the most important value of weak acid pH distribution systems. A solution is buffered when it is within 1.0 pH unit of a pK_a for a weak acid—be it mono-, di-, or tri-protic acid. Thus, the acidification of our oceans, buffering in the blood, exercise physiology, and brewing good beer are all dependent on control of pH in weak acid systems.

Sec 5.8 Technical References

These are intended mostly for your chemistry professor's use.

[1] Falkowski, P. et al. (2000) The Global Carbon Cycle: A test of our knowledge of Earth as a system. *Science*, 290 (5490), 291–296.

[2] Choi, Y. S. et al. (2002) Determination of oceanic carbon dioxide using a carbonate-selective electrode. *Anal. Chem.* **2002,** *74*, 2435–2440.

[3] Hardege, J. D. et al. (2011). Analytical challenges and the development of biomarkers to measure and to monitor the effects of ocean acidification. *Trends in Anal. Chem.*, Vol. *30* (8), 1320–1326.

[4] Palmer, J., & Kaminski, C. (2013). Water: A comprehensive guide for brewers. Brewers Publication [www.BrewersAssociation.org]

[5] Messonnier, L., Kristensen, M., Juel, C., & Christian Denis, C. (2007). Importance of pH regulation and lactate/H+ transport capacity for work production during supramaximal exercise in humans. *J. Applied Physiol. 102*(5), 1936–1944.

UNIT 6
Solubility of Aqueous Salts

WHY STUDY THIS UNIT?

The goal of this unit is to help you develop a qualitative understanding of how "insoluble" salts partially dissolve in water. Specifically, this unit will cover how to . . .

- Visualize the solubility product constant, $\mathbf{K_{sp}}$, for a slightly soluble salt;
- Determine the relationship between the non-equilibrium constant, $\mathbf{Q_{sp}}$, and the equilibrium constant, $\mathbf{K_{sp}}$;
- Visualize Chemical Potential Energy of K_{sp} vs. Q_{sp} (Figure 6.4) so you can truly understand the solubility of slightly soluble salts;
- Use the Q_{sp} to K_{sp} ratio to predict if a precipitate will form when the two product ions are mixed together.

Most textbooks devote only part of a chapter on the topic of **solubility equilibria**, (usually in Chapter 16 or 17).

Sec 6.1 Introduction

In your first semester chemistry course, you probably recall the term "insoluble salts" and you may have memorized the "solubility rules" for salts (ionic compounds) in water. In this course, chemists refer to "insoluble salts" as "slightly soluble salts" because some of the salt particles dissolve in water. How do scientists know that a tiny amount of a slightly soluble salt dissolves in water?

Let's use barium sulfate, $BaSO_4$, as an example of a slightly soluble salt. First, if you were trying to dissolve $BaSO_4$ in water, you might place say 10.0000 grams of it in a 1.00 L (weigh the empty flask first) and then add 1,000 mL of pure water. Stir vigorously to try to dissolve this salt. Next, to filter the solution, place filter paper (also pre-weighed) on a funnel (Buchner funnel would give better results than a glass funnel), and filter it into a pre-weighed filter flask (Erlenmeyer flask with a side arm). Evaporate the water from the solution by carefully heating it until all of the water has evaporated. Finally, re-weigh the filter flask. How much solid $BaSO_4$ would there be in the flask? The answer would be about 0.0023 grams. To verify that most of the salt did not dissolve, dry the filter paper, which collected the undissolved salt. To find this mass, subtract the weight of the filter paper (which was pre-weighed) and you would get an answer of 9.9977 g. In other words, almost all of this salt did not dissolve in 1.000 L of water—only 0.0023 grams.

The above filtration experiment is an "old fashioned" way to determine how much of the $BaSO_4$ dissolved in 1.00 L of water. A more modern way would be to determine the conductivity for pure water and then for the filtered solution. Conductivity is the best way to determine how pure your water is. Its conductivity is about 5.5 * 10^{-8} Siemens per centimeter (5.5 * 10^{-8} S/cm). Why does pure water conduct? You should already know this answer from the previous unit (Unit 5). Pure water is very, very slightly ionized: 1.0 * 10^{-7} M (or 1 in 10 million water molecules have ionized into H^+ and OH^-). Next, we can determine the conductivity of the barium sulfate solution from the above experiment. It is about 3.1 * 10^{-6} S/cm[1], which is about 100X greater than the conductivity of pure water. This is indeed a tiny number. How could we compare it with something more familiar? If you dissolve 0.64 mg of NaCl (0.00064 g) in 1.000 L of water, then check its conductivity—you get an answer of about 1.0 * 10^{-6} S/cm. Therefore, only a very tiny amount of $BaSO_4$ dissolved in 1.0 L of water.

So in the real world outside the classroom—Who on Earth would want to know that $BaSO_4$ is essentially insoluble in water? One answer is anyone who collects minerals—$BaSO_4$ is barite, and it forms a mineral that looks like a rose—the barite rose (see Figure 6.1 below). Another answer is that petroleum engineers who are working in an oil field have found that barium sulfate in the processing water plugs up the production lines, which are used to transport oil to the oil refinery plant[2]. Think like a business major—less oil flow means less oil/gasoline and hence less profits for Exxon (in 2014 their profit was \$40.6B on annual sales of \$482.3B).

Returning to chemistry—the chemical reaction for this slightly soluble salt is as follows:

EQ 6.1: $Ba^{2+}(aq) + SO_4^{2-}(aq) \rightarrow BaSO_4(s)$ {the net ionic equation}

In other words, this precipitation reaction occurs when an oil company pumps seawater with its high sulfate ion concentration into hard water (containing Ba^{2+}, Ca^2, and Sr^{2+}) in the reservoir near the oil deposits. Consequently, barium sulfate precipitates out and clogs up the pipeline. Ironically, the oil and gas industry uses barite (barium sulfate) in its drilling mud[3]. Also, if you are a pre-med major—before undergoing an X-ray examination of the gastro-intestinal tract, the patient ingests barium sulfate so their GI tract can be analyzed.

Figure 6.1 Photograph of a barite rose, $BaSO_4$ (s)

©Citrina/Shutterstock.com

Sec 6.2 K_{sp} for a 1:1 Cation:Anion System

Overall, these slightly soluble salts when mixed with water undergo a reversible reaction. Does this mean that they begin with equilibrium when they are mixed?

EQ 6.2: $BaSO_4(s) \leftrightarrow Ba^{2+}(aq) + SO_4^{2-}(aq)$

The answer is "no"! At first, there is 100% solid $BaSO_4$ and only the forward reaction can occur. This is a *non-equilibrium condition* because the rate of the forward process is much greater than the rate of the backward process.

EQ 6.1: $BaSO_4(s) \rightarrow Ba^{2+}(aq) + SO_4^{2-}(aq)$

As this reaction continues, enough ions accumulate such that the backward process begins to occur, and these ions precipitate $BaSO_4$ (s). In other words, the forward process will be greater than the backward process. That is, the rate of dissolving is greater than the rate of precipitation. Next, the two rates will eventually become equal to each other as the reversible reaction attains equilibrium:

EQ 6.2: $BaSO_4(s) \leftrightarrow Ba^{2+}(aq) + SO_4^{2-}(aq)$

So for every $BaSO_4$ (s) that dissolves into its ions, Ba^{2+} and SO_4^{2-} will collide with each other and precipitate out of solution.

Suppose that we have 1.00 L of aqueous solution with an excess of $BaSO_4$ precipitate. Does it matter how much solid we have under the equilibrium condition? The answer may surprise you: as long as there is enough solid to undergo this "exchange," then it does not matter whether there is a ton of it (worth $25/ton) or just a tiny amount. Do you recognize EQ 6.2 as a heterogeneous, reversible reaction? First, recall that only the most fluid state (aqueous) is included in the equilibrium expression:

$$K_{sp} = [Ba^{2+}]_{equil} * [SO_4^{2-}]_{equil} = 1.0 * 10^{-5} \text{ at } 25°C \text{ (room temperature)}$$

How was this specific K_{sp} value obtained? The empirical answer (i.e., knowledge source from experiments) is that the gram-solubility, 0.0023 g/L, was determined as described in the above section. Next, you can calculate the molar-solubility using the molar mass (MM = 233.4 g/mol) as follows:

$$? \text{ mol/L} = (0.0023 \text{ g})/(233.4 \text{ g/mol}) = 1.05 * 10^{-5} \text{ mol/L} = 1.0 * 10^{-5} \text{ M}$$

Finally, how would you calculate the K_{sp} value at 25°C (this was the experimental temperature)? The best approach is to use a symbol for the molar solubility: let s = molar solubility; then solve the K_{sp} expression in terms of this variable:

$$K_{sp} = [Ba^{2+}]_{equil} * [SO_4^{2-}]_{equil} = s * s \text{ and } \mathbf{K_{sp} = s^2}$$

With this equation you can solve two types of problems:

1. Given the value of s, solve for K_{sp}, and
2. Given the value of K_{sp}, solve for s

Thus, $s = 1.0 * 10^{-5}$ M, and $K_{sp} = s^2 = (1.0 * 10^{-5})^2$ so, $\mathbf{K_{sp} = 1.0 * 10^{-10}}$

You may wonder about the meaning of this very small number for the molar solubility of $BaSO_4$. What the heck does it mean? Suppose we have one mole of $BaSO_4$ (s), how many moles of Ba^{2+} and SO_4^{2-} ions would we get?

EQ 6.2: $BaSO_4 (s) \leftrightarrow Ba^{2+} (aq) + SO_4^{2-} (aq)$

Assume: 1 mole → $1.0 * 10^{-5}$ M & $1.0 * 10^{-5}$ M at equilibrium

How can you make sense out of these calculations, which yield only 10^{-5} moles? Recall that one mole is just a particular count of particles; thus, only one out of 100,000 ion-pairs would dissolve to produce one Ba^{2+} and one SO_4^{2-} ion. This is like finding one person at random out of a city of 100,000.

Sec 6.3 K_{sp} for a 3:2 Cation:Anion System

Calcium phosphate, $Ca_3(PO_4)_2$, is another slightly soluble salt and its K_{sp} is more complex than the one for $BaSO_4$. First, where is $Ca_3(PO_4)_2$ found in the real world? It is in cow's milk, and it is the principal component of bone mineral. It is the primary chemical in the mineral apatite (see Figure 6.2). Apatite is one of only a few minerals that is produced and used by biological systems. In agriculture it is used as a source of phosphorus fertilizer.

Figure 6.2 Apatite mineral is $Ca(Ca_3(PO_4)_2)_6X_2$, where X = F^-, Cl^-, or OH^-

Let's get back to chemistry: What is the relationship between the molar solubility, *s*, and K_{sp} for this salt?

EQ 6.3: $Ca_3(PO_4)_2 (s) \leftrightarrow 3\ Ca^{2+} (aq) + 2\ PO_4^{3-} (aq)$ $K_{sp} = 2.0 * 10^{-29}$ at 25°C

To make the calculations as simple as possible: Suppose that 1 mole of calcium phosphate is mixed with water—

- ❍ How many moles of each ion are present at equilibrium?
- ❍ What is the molar solubility of $Ca_3(PO_4)_2$?

EQ 6.3: $Ca_3(PO_4)_2$ (s) ⟷ 3 Ca^{2+} (aq) + 2 PO_4^{3-} (aq) $K_{sp} = 2.0 * 10^{-29}$ at 25°C

Assume: 1 mole ➔ 3 *s* + 2 s

- ❍ What is the molar solubility of this salt at equilibrium? s = ?
- ❍ What is the concentration of calcium ion? $[Ca^{2+}]_{equil}$ = ? Answer: 3 *s*
- ❍ What is the concentration of phosphate ions? $[PO_4^{3-}]_{equil}$ = ? Answer: 2 *s*
- ❍ What is the relationship between molar solubility, *s*, and K_{sp}?

$$K_{sp} = [Ca^{2+}]^3_{equil} * [PO_4^{3-}]^2_{equil} = [3s]^3 * [2s]^2 = [3*3*3\ s^3] * [2 * 2\ s^2] = 27s^3 * 4s^2$$

$$\mathbf{K_{sp} = 108\ s^5}$$

Given the value of the solubility constant, $K_{sp} = 2.0 * 10^{-29}$, what is the molar solubility, *s*, of calcium phosphate?

- ❍ Solving for *s*: $s = (K_{sp}/108)^{1/5} = ?$

Please note that the fraction 1/5 is equal to 0.20, so . . .

$$s = (K_{sp}/108)^{0.20} = (2.0 * 10^{-29} /108)^{0.20} = (1.85 * 10^{-31})^{0.20} = 7.14 * 10^{-7}$$

- ❍ What are the units on s? Always: mol/L = M, thus s = **$7.14 * 10^{-7}$ M**
- ❍ What is $[Ca^{2+}]_{equil}$? Answer: 3 *s* = 3 * 7.14 * 10^{-7} M = **$2.1 * 10^{-6}$ M**
- ❍ What is $[PO_4^{3-}]_{equil}$? Answer: 2 *s* = 2 * 7.14 * 10^{-7} M = **$1.4 * 10^{-6}$ M**

Hopefully, you recognize the equation that leads to a successful solution:

$$K_{sp} = [Ca^{2+}]^3_{equil} * [PO_4^{3-}]^2_{equil} = [3s]^3 * [2s]^2 \qquad \textbf{so, } \mathbf{K_{sp} = 108\ s^5}$$

Please note that with this equation you can solve for the unknown for <u>two</u> types of problems:

Given *s*, solve for K_{sp}
Given K*sp*, solve for *s*

This is the efficient, reliable way to work these problems.

Conversely, a poor method is to just plug numbers into your calculator. Why? If you happen to invert two numbers—say you key in A/B when it should be B/A, then you will get the wrong answer. Also, it is too easy to forget one step out of a multi-step problem. Also, you may forget to multiply molar solubility, *s*, by two in the expression: $[2s]^2$ and your answer is off by a factor of 4X (2^2)—a 400% error. *Word to the Wise*: Always set up the algebraic expression ($K_{sp} = 108\ s^2$ for EQ 6.3) before you write numbers in the equation and plug them in your calculator.

Sec 6.4 Q_{sp}: Non-Equilibrium Solubility of a Slightly Soluble Salt

If you have several aqueous solutions of soluble anions and metal cations, how can you pick a chemical reaction that will form a precipitate? Let's use calcium fluoride, CaF_2, as our example. CaF_2 is the primary chemical in *fluorite*, which is a colorful mineral (see Figure 6.3). It is used in excimer lasers and in the manufacture of optical components such as windows and lenses. Suppose you are in the lab, and your goal is to precipitate out CaF_2 (s). How would you do this reaction in the lab?

Figure 6.3 Pure Fluorite is CaF_2 (s)—a colorless crystal; however, small amounts of impurities give it different colors.
©Nastya Prineva/Shutterstock.com

Answer: mix soluble Ca^{2+} solution with a soluble F^- solution as follows:

EQ 6.4: $$Ca^{2+}(aq) + 2\ F^-(aq) \rightarrow CaF_2(s)$$

Look at the balanced equation in EQ 6.4 and think "stoichiometry." You should see how to make the solid precipitate: Mix one mole of soluble calcium ion with two moles of soluble fluoride ion. Let's say you decide to have 1.00 L of solution. What would happen? Would you see a precipitate? Well, let's do some calculations:

Find $Ca(NO_3)_2$ (s) in the chemical stockroom because it is a soluble salt of calcium (recall that "all metal nitrate salts are soluble in water").

- Calculate the molar mass of $Ca(NO_3)_2$: *MM* = 164.088 g/mol.
- Since the goal is to have 1.00 Liter after the two solutions are mixed, then each solution should have a volume of 500.0 mL of this solution.
- Weigh out 82.044 g of the salt (1/2 of *MM*) and dissolve it in enough water to make 500 mL of solution.

How many grams of sodium fluoride (MM = 41.989 g/mol) are needed?

- NaF (aq): 1.0 M = 41.989 g/L so for 500.0 mL (1/2 of 1 L): dissolve 20.994 g NaF in enough water to make 500.0 mL.
- Mix these two aqueous solutions together to form the CaF_2 precipitate:

EQ 6.4: Ca^{2+} (aq) + 2 F^- (aq) → CaF_2 (s)

Molar mix:	1.0000 M + 1.0000 M	→ ??
Stoichiometry:	1 mol Ca^{2+} + 2 mol F^-	→
Moles available:	0.5000 mol + 0.5000 mol	
Moles reaction:	0.2500 mol + 0.5000 mol	→ 0.2500 mol
Ca^{2+} is excessive	<u>F– is the limiting reactant</u>	

- Concentration of CaF_2: ?M = 0.2500 mol/1.0000 L = **0.2500 M**

  <u>Ooops</u>—we forgot to check the solubility product of CaF_2: $K_{sp} = 3.9 * 10^{-11}$

  So what do we do with K_{sp}? Answer:

- First, find molar solubility, *s*, where **s** = molar solubility of CaF_2
- Then, solve for K_{sp} as follows:

EQ 6.4: Ca^{2+} (aq) + 2 F^- (aq) → CaF_2 (s)

EQ 6.5: Ca^{2+} (aq) + 2 F^- (aq) ← CaF_2 (s) {think *Right to Left*}

s + 2*s*

(Continued)

❍ $K_{sp} = [Ca^{2+}]_{equil} * [F-]^2_{equil} = [s] * [2s]^2 = 4s^2 = 3.9 * 10^{-11}$

So, $s = ((3.9 * 10^{-11})/4)^{1/2} = (9.8 * 10^{-12})^{0.5} = 3.1 * 10 - 6 =$ **$3.1 * 10^{-6}$ M CaF_2**

How many moles of CaF_2 were actually *dissolved* in our reaction?

? mol = 1.0000 L * $3.1 * 10^{-6}$ mol/L = $3.1 * 10^{-6}$ moles in solution

Please compare this correct answer, $3.1 * 10^{-6}$ moles with the incorrect answer that was determined using simple stoichiometry (0.2500 moles).

So, how can these *non-equilibirum calculations* be simplified? Answer: To simplify these calculations, chemists have developed and used a "non-equilibrium constant" termed **Q_{sp}**. The mathematics of Q_{sp} calculations are identical to that for K_{sp}. Technically, Q_{sp} is called the *ion product*. For the reaction above,

EQ 6.5: $$Ca^{2+} (aq) + 2 F^- (aq) \leftrightarrow CaF_2 (s)$$

$$Q_{sp} = [Ca^{2+}]_{initial} * [F-]^2_{initial}$$

How will this concept of ion product, Q_{sp}, help us simplify the above situation?

The answer is that by comparing the numerical values of Q_{sp} with K_{sp}, we can determine whether or not a precipitate will form when the two solutions, soluble cation and soluble anion, are mixed together.

❍ If **$Q_{sp} > K_{sp}$**, then a *precipitate forms and its solution is saturated.* Here is how you should view this situation: Qsp > Ksp means there are more ions in solution than there are in the equilibrium solution. Hence, a precipitate forms and its solution will be saturated with the two ions.

❍ If **$Q_{sp} < K_{sp}$**, then there is *no precipitate*, and the salt *solution is unsaturated.* In other words, there are not enough ions in solution to effect precipitation.

❍ If **$Q_{sp} = K_{sp}$**, then it is a *saturated solution at equilibrium* and no precipitate forms.

Let's work this example problem (see boxed text above) using these criteria.

Given: $[Ca^{2+}]_{initial} = 0.5000$ M and $[F^-]_{initial} = 0.5000$ M, and $K_{sp} = 3.9 * 10^{-11}$

EQ 6.5: $$Ca^{2+} (aq) + 2 F^- (aq) \leftrightarrow CaF_2 (s)$$

$Q_{sp} = [Ca^{2+}]_{initial} * [F-]^2_{initial}$

$Q_{sp} = [Ca^{2+}]_{initial} [F^-]^2_{initial} = [0.5000] * [0.5000]^2 = 0.125 = 1.25 * 10^{-1}$

$Q_{sp} > K_{sp}$, thus CaF_2 precipitates and the solution is saturated with Ca^{2+} & F^- ions.

How is it possible to mix Ca^{2+} and F^- (aq) ions and <u>not</u> form a precipitate?

Given: $[Ca^{2+}]_{initial} = 0.0100$ M and $[F^-]_{initial} = 1.00 * 10^{-5}$ M, and $K_{sp} = 3.9 * 10^{-11}$

EQ 6.5: $Ca^{2+}(aq) + 2 F^-(aq) \longleftrightarrow CaF_2(s)$

$Q_{sp} = [Ca^{2+}]_{initial} * [F^-]^2_{initial}$

$Q_{sp} = [Ca^{2+}]_{initial} [F^-]^2_{initial} = [0.0100]*[1.00 * 10^{-5}]^2 = [1.00 * 10^{-2}]*[1.00 * 10^{-10}]$

$Q_{sp} = 1.00 * 10^{-12}$

$Q_{sp} < K_{sp}$, thus <u>no</u> precipitate forms and the solution is <u>unsaturated.</u>

Sec 6.5 Chemical Potential Energy: K_{sp} vs. Q_{sp}

When Q_{sp} is <u>not</u> equal to K_{sp}, what determines the direction (R → P or P → R) of a reversible reaction? The underlying concept that determines the answer to this question is—the relative *chemical potential energies*. This relationship is shown in a diagram (see Figure 6.4) of chemical PE (y-axis) versus relative concentrations of product ions with respect to their equilibrium concentration (x-axis).

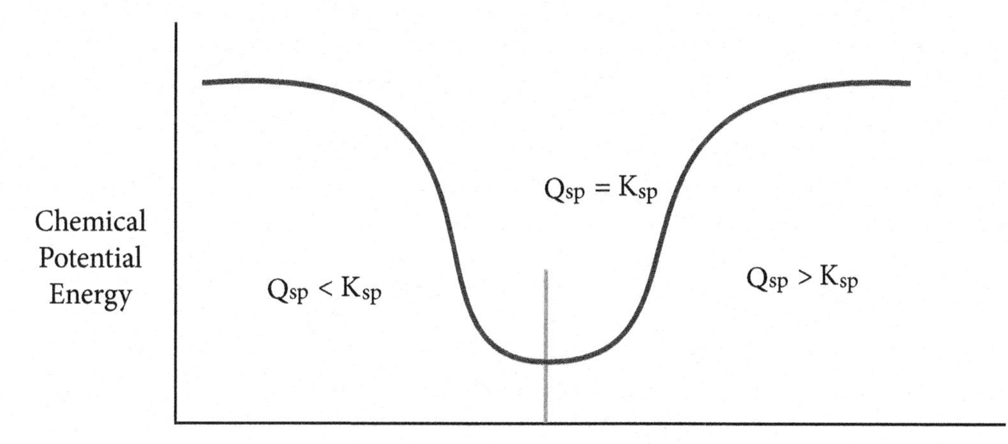

Figure 6.4 The relative chemical potential energy of different values of the ion product, Q_{sp}, with respect to K_{sp}

Please note that there are three distinctive regions in Figure 6.4: When Q_{sp} is less than K_{sp}, there is a deficiency of product ions (i.e., [Cation] & [Anion] in Q_{sp}) with respect to the equilibrium concentration of ions as expressed by K_{sp}. Conversely, when Q_{sp} is greater than K_{sp}, there is an excess of product ions with respect to the equilibrium concentrations of ions. You should see that both of these non-equilibrium conditions have higher chemical potential energy than the chem PE of the equilibrium constant. Also, note that the lowest chemical potential energy in Figure 6.4 is where $Q_{sp} = K_{sp}$. This means that *the most stable state* for this chemical system is when the ion product is equal to K_{sp}. Consequently, if $Q_{sp} < K_{sp}$, then there is a *deficiency of product ions*, and the system shifts to the right (i.e., $Rate_{dissolving} < Rate_{ppt}$) until the equilibrium condition is restored. Likewise, if $Q_{sp} > K_{sp}$, then there is an *excess of product ions*, and the system shifts to the left (i.e., $Rate_{ppt} > Rate_{dissolving}$) to reduce the ion product. This process continues until the equilibrium condition is satisfied with $Q_{sp} = K_{sp}$. Overall, the ion product Q_{sp} will shift until it reaches the value of the equilibrium constant, K_{sp}. Do you understand this relationship? If "yes," then you understand the chemistry of slightly soluble salts!

Sec 6.6 Predicting Whether Precipitation Will Occur

In all of the above sections, it is assumed that to determine K_{sp} you mix two ions, say Pb^{2+} and CO_3^{2-} (aq), that form a slightly soluble salt, then measure the amounts of dissolved ions. The quantitative value of knowing this equilibrium constant is that you can determine whether or not a precipitate will form when two particular concentrations of ions are mixed together.

Conversely, what if you want to dissolve a precipitate—What should you do? The answer is to apply *LeChatlier's principle* by finding a reagent that will remove ions (e.g., Pb^{2+} and CO_3^{2-} (aq)) from the solution and hence shift the equilibrium condition to the right. This "removal of ions" creates a stress with respect to K_{sp}, and the system responds by removing the stress. What is the stress in this case?

EQ 6.6: $PbCO_3\,(s) \leftrightarrow Pb^{2+}\,(aq) + CO_3^{2-}\,(aq)$

In general—the easiest way to remove one of these ions is to use a reagent that will react to form a molecular species, and hence remove one of the ions. Specifically, you know that carbonic acid is a weak acid in water, which is mostly molecular rather than dissolved as ions. Thus, you should add strong acid, say HCl (aq), to the chemical system in EQ 6.6. Hydrochloric acid is ~100% ionized as H_3O^+ (or H^+ in a net ionic equation) so its reaction with carbonate ion essentially "goes to completion" (i.e., an irreversible reaction). Next, carbonic acid decomposes to form water and carbon dioxide gas, which escapes from the system as shown in EQ 6.7 below:

EQ 6.7: $2\,H^+\,(aq) + CO_3^{2-}\,(aq) \rightarrow H_2CO_3\,(aq) \rightarrow H_2O\,(l) + CO_2\,(g)$

How can you show how this reaction, EQ 6.7, affects the $PbCO_3$ chemical system? Answer: Add these two equations together to get the net ionic equation, EQ 6.8:

EQ 6.6: $PbCO_3\,(s) \leftrightarrow Pb^{2+}\,(aq) + CO_3^{2-}\,(aq)$

EQ 6.7: $2\,H^+\,(aq) + CO_3^{2-}\,(aq) \rightarrow H_2O\,(l) + CO_2\,(g)$

EQ 6.8: $PbCO_3\,(s) + 2\,H^+\,(aq) \rightarrow Pb^{2+}\,(aq) + H_2O\,(l) + CO_2\,(g)$

In EQ 6.8 the precipitate is a reactant and it is dissolved on the product side, which has soluble lead (II) ion and two molecular species—water and carbon dioxide gas. In general, if the anion of a slightly soluble salt forms a weak acid, then addition of a strong acid will remove that anion in its acidic form as a weak acid. Now, you know how to dissolve a precipitate!

Sec 6.7 Summary

In this unit, we discussed how to determine the solubility of slightly soluble salts in aqueous systems. You should remember how the concept of the equilibrium constant, K_{sp}, was determined. First, measure how much of the salt dissolved in grams per Liter, then how to convert to *s*, molar solubility (mol/L). Next, please get in the habit of solving for K_{sp} in terms of *s* before you substitute in the numbers and perform the math operations on your calculator. Also, remember that all K_{sp} problems involve reversible reactions, ⟷, that often are under non-equilibrium conditions (i.e., Q_{sp} calculations). To understand the relationship between K_{sp} and Q_{sp}, you need to picture Figure 6.4 where there are three regions: $Q_{sp} < K_{sp}$, $Q_{sp} = K_{sp}$, and $Q_{sp} > K_{sp}$. In all three regions, the fundamental question is . . . What is the ratio of product ions to equilibrium ions? The answer is that the system has "depleted ions" if $Q_{sp} < K_{sp}$; "equal ratio of ions" if $Q_{sp} = K_{sp}$; and "excessive ions" if $Q_{sp} > K_{sp}$. Both non-equilibrium conditions, $Q_{sp} \neq K_{sp}$, have higher chemical potential energies than the *more stable equilibrium condition*, where $Q_{sp} = K_{sp}$. Finally, what is needed to dissolve a precipitate? Answer: find a reagent to remove one of the product ions by transforming it into its molecular form.

Sec 6.8 Technical References

[1] Rosseinsky, D. R. (1958). The solubilities of sparingly soluble salts in water: Part 5—The solubility of barium sulphate at 25°C. *Trans. Faraday Soc., 54*, 116–118.

[2] Jin, H., Yang, B., Yang, S., & He, G. (2015). An experimental and modeling study of barite deposition in one-dimensional tubes. *J. Petroleum Sci. Engr., 125*, 107–116.

[3] Hanor, J. (2000). Barite–Celestine Geochemistry and Environments of Formation *Rev. Mineralogy Geochem., 40*, 193–275.

UNIT 7

Thermodynamic Systems

WHY STUDY THIS UNIT?

The goal of this unit is to help you develop a qualitative understanding of how to predict when a chemical reaction is spontaneous. Specifically, this unit will cover how to . . .

- ❍ Understand that entropy means "dispersal of molecules" as a thermodynamic variable;
- ❍ Visualize entropy as "increase in the number of microstates";
- ❍ Show how spontaneous chemical reactions are dependent on Gibbs free energy, enthalpy, entropy, and temperature;
- ❍ Understand how Gibbs free energy is dependent on the *chemical potential energy* of chemical bonds;
- ❍ Understand the four different types of thermodynamic reactions.

Most textbooks devote one chapter on **thermochemistry** (in first semester) and one on **thermodynamics** (usually Chapter 18 or 19).

Sec 7.1 Introduction

In your first semester chemistry course, you probably learned how to calculate the change in enthalpy, ΔH_{rxn}, for a chemical reaction. For these calculations, you are given the balanced equation and the enthalpy of formation, ΔH_f, for each chemical species (molecule, ion, or atom). Next, remember how you interpreted the answer? If ΔH_{rxn} is negative, then reaction is *exothermic* and heat is given off by the system. In other words, heat is lost by the system—hence, the negative sign—and that heat is absorbed by the surroundings, which get hotter. Conversely, if ΔH_{rxn} is positive, then the reaction is *endothermic* and heat is absorbed by the system; hence, the positive sign. In other words, heat flow: surroundings → system, which means that the system sucked heat out of its surroundings, which get cooler.

Your chemistry instructor may have said that "exothermic reactions tend to be spontaneous," while "endothermic reactions tend to be non-spontaneous." Also, to make an endothermic reaction occur—the system has to be heated continuously. Stop heating and the reaction stops immediately. Do the words "tends to be" bother you? If "yes," that means you are thinking about it (rather than just memorizing the material). In this unit, we will replace this tentative language with more assertive language.

We will introduce a new thermochemical term called "entropy." You may have heard about it elsewhere (like in a biology class)—entropy is amount of disorder in a system. Unfortunately, this association (entropy = disorder) is not the best way to describe entropy. Entropy is best thought of as "*dispersal of molecules*" (ions or atoms). Dispersal means "spreading out," and entropy is the effect this dispersal has on energy in a physical process (e.g., H_2O (l) → H_2O (g)) or a chemical reaction. The best way to illustrate entropy is explored in the next section.

Sec 7.2 Microstates and Entropy

The concept of entropy can be related to the number of possible "microstates" for a given chemical species. If you have a pure substance (element or compound), then its gas phase has many more microstates than its liquid phase, which in turn has more microstates than its solid phase. To understand microstates in the three phases of matter, let's try to predict where the molecules (atoms or ions) are located in each phase. For the solid phase (Figure 7.1a), where would you put the "missing" molecule? Obviously, there is only one choice; that is, there is only one microstate for a solid that has uniform distribution of rows and columns of molecules/ions (like metal atoms). When there are fewer microstates, then there is less entropy (similar to amount of disorder). In the liquid phase (Figure 7.1b), suppose these circles represent water molecules—recall that in liquid water, three water molecules are hydrogen-bonded together. Which direction is each chain of three molecules moving? The answer is that this cannot be determined. For example, if you say the three near the top on a diagonal

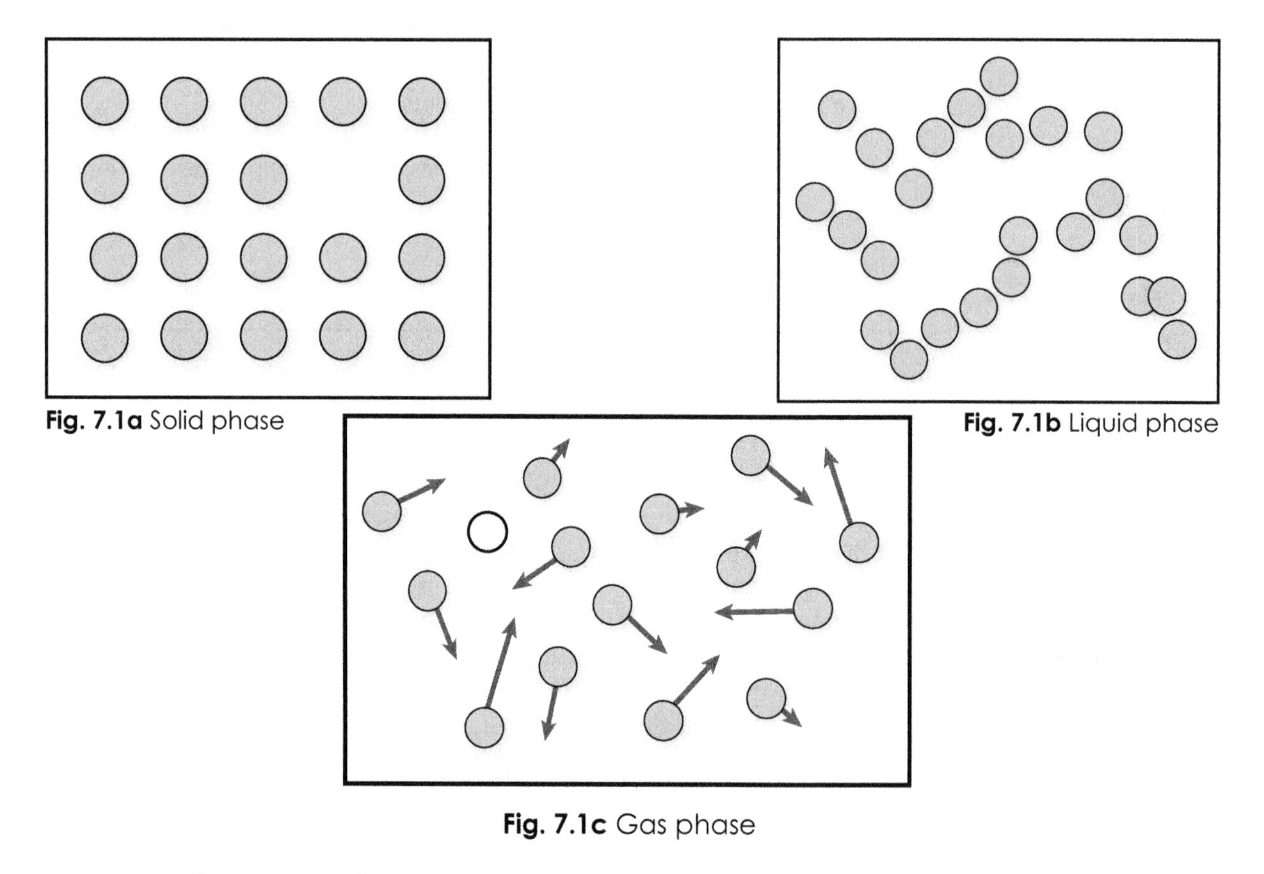

Fig. 7.1a Solid phase

Fig. 7.1b Liquid phase

Fig. 7.1c Gas phase

Figure 7.1 Microstates in solids (7.1a), liquids (7.1b), and gases (7.1c)

are moving "up," then the odds are just as good that they are moving "down." So each chain of circulating molecules is moving independently of the other chains. This variety of possibilities means that there are many more microstates in a liquid as compared to its solid state. In the gas phase (Figure 7.1c), each molecule is separated by a huge amount of empty space and each one tends to have a different velocity (the length of the arrow—short or long). In this diagram, can you predict the velocity of the molecule shown as an open circle "o"? Is this molecule moving up with a fast velocity, down with a fast velocity, to the left/right, up or down with a slow velocity? The answer is that it could be any one of these possibilities; hence, there are many, many microstates for molecules in the gas phase.

Let's discuss the number of microstates for molecules in a gas in more detail. What if we knew there was another molecule in Figure 7.1c, but we did not know where it was located? Could we figure it out logically? No! . . . it could be anywhere in this diagram. Try this analogy: Paste this diagram (Figure 7.1c) on the wall (like a dart board), step away about ten feet and throw a dart at it (Figure 7.2). The "missing molecule" could be anywhere the dart lands on the board. You might object to taking these "random shots" at determining the location of a gas molecule, but there is no organized distribution of molecules and their velocities in the gas phase. Furthermore, recall from chemistry (last semester) the kinetic molecular theory of gases: (1) the volume of each molecule is so small as compared to the huge volume of empty space surrounding it—that its volume can be zero—just an infinitesimally small dot; (2) each molecule is moving in rapid, random directions; (3) the collisions between molecules are elastic (i.e., no energy is lost); and (4) pressure is due to collisions between molecules and the walls of the container. This discussion shows that it is virtually impossible to predict the location and velocity of individual molecules in the gas phase. Furthermore, when there are millions and billions of gas molecules—the number of possible microstates approaches "infinity."

Another way to understand the relationship between number of possible microstates and entropy is to imagine two bulb reservoirs linked together by a glass tube as shown in Figure 7.3. In this example, there are four molecules (A, B, C, and D) in one bulb and none in the other. In Figure 7.3, this arrangement of molecules is shown in sketch 4t: 0b, where t = top bulb and b = bottom bulb. How long does this particular distribution (ratio of four to zero molecules) last? The answer is not very

Figure 7.2 Darts and a dart board

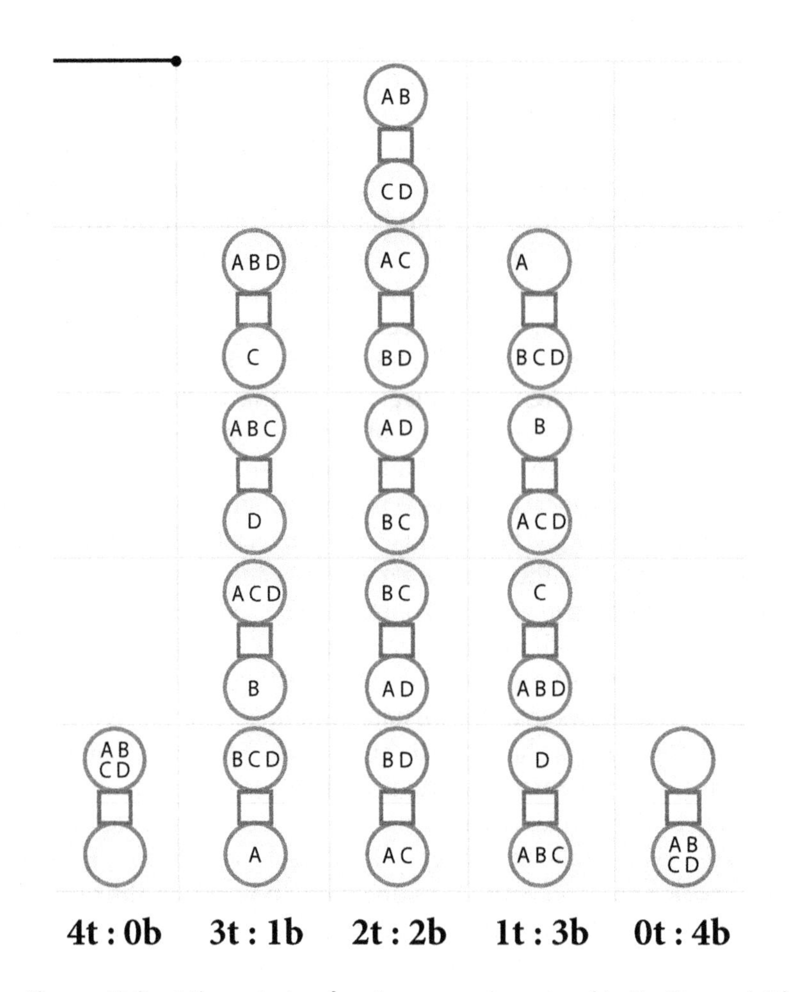

Figure 7.3 Microstates for 4 gas molecules (A, B, C, and D) in two interconnected bulbs (t = top and b = bottom).

long because of the rapid, random motion of gas molecules and "diffusion." As shown in Figure 7.3, the odds of finding all four molecules in just one bulb with a vacuum in the other (4t:0b & 0t:4b) are two in fifteen (13% of the time). Next, note that the molecules are more evenly distributed in the other microstates (3t:1b, 2t:2b, & 1t:3b), which occurs in thirteen of fifteen, or 87% of the time.

Furthermore, in a real-world system found in a laboratory, if a piece of glassware has two small glass bulbs connected by a very thin capillary tube, then this would contain many, many molecules—say, $6.0 * 10^{16}$ molecules. What would be the odds that there would be only 10,000 molecules in one flask and $6.0 * 10^{12}$ in the other? Answer: the odds would be one out of 1,000 billion that this "uneven distribution" would occur. Consequently, the odds of finding a vacuum in one flask and $6.0 * 10^{16}$ molecules in the other are much, much smaller—infinitesimally small—essentially impossible.

Entropy is the change from an initial state with fewer microstates to a final state that has more microstates. So referring to Figure 7.3, if we start with a system that has only one microstate (e.g., 4t:Ob), then the chemical system will spontaneously change to one with four microstates (like 3t:1b), and then this intermediate state will change to the final state with five microstates (e.g., 2t:2b). This final state will not spontaneously change to one with fewer microstates (e.g., 1t:3b or 0t:4b).

What would be another example of this spontaneous process, which occurs in the real world? Suppose we have glassware that has two 1-Liter bulbs connected by a glass tube that has a two-way valve. This valve is <u>closed</u> when the valve handle is perpendicular _||_ to the tube, and <u>open</u> when the valve is parallel to the tube _=_. If Bulb A is full (containing billions and billions of molecules) and Bulb B contains a vacuum—when the valve is open, molecules will spontaneously flow from the gas-filled to the empty bulb. This process is called *diffusion*, and it is an example of entropy. While the valve was closed, there was only one microstate (full:empty). When the valve is opened, the gas molecules diffuse into the empty bulb and this diffusion process contains many, many successive microstates (similar to going from 4t:0b to 3t:1b to 2t:2b in Figure 7.3). When will this spontaneous process (full to empty) stop? Answer: when both bulbs have the same number of microstates, then the system has reached "equilibrium" (Rate of left-to-right flow equals rate of right-to-left flow) between the two bulbs.

When a gas in a sealed vessel is heated, there is a spontaneous change from fewer microstates to more microstates (Figure 7.4). That is, there are fewer kinetic energy microstates (five different KE's in Figure 7.4) at the lower temperature. The system is heated and it spontaneously dissipates the thermal energy into more microstates (nine different KE's). Cooling this gas back to the lower temperature would be non-spontaneous because energy must be dissipated from the system to the surroundings.

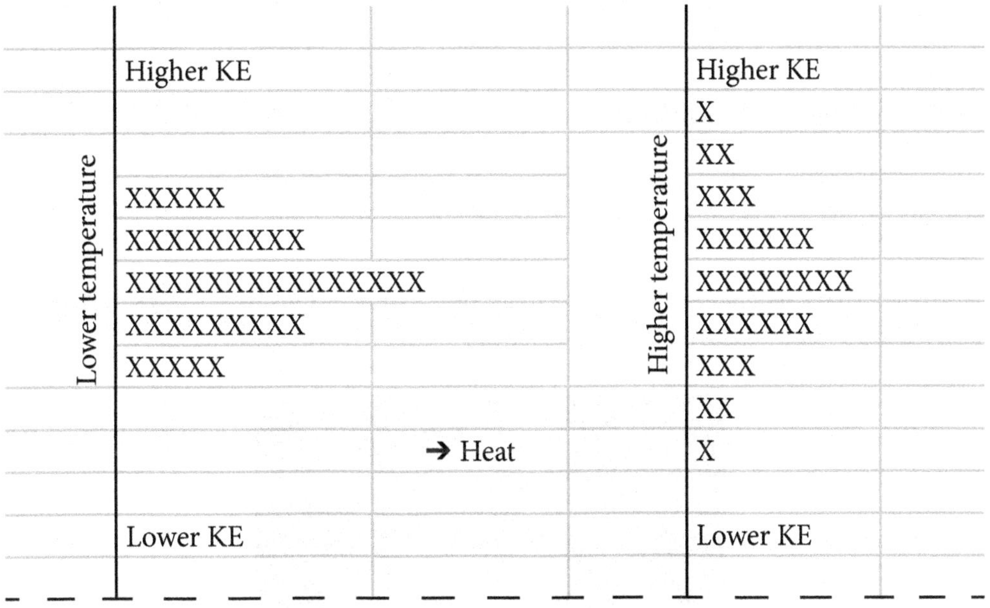

Lower temperature		Higher temperature
Higher KE		Higher KE
		X
		XX
XXXXX		XXX
XXXXXXXXX		XXXXXX
XXXXXXXXXXXXXXX		XXXXXXXX
XXXXXXXXX		XXXXXX
XXXXX		XXX
		XX
	→ Heat	X
Lower KE		Lower KE

Figure 7.4 A gas at a lower temperature has fewer microstates (five microstates in this example) is heated to a higher temperature where there are more microstates (nine microstates); where x = molecules

Sec 7.3 When Is an Exothermic Reaction Spontaneous?

Now we are ready to replace the tentative phrase "tends to be spontaneous" with the more assertive phrase "spontaneous at temperature, T." First, you must understand the concepts of Gibbs free energy and change in Gibbs free energy, ΔG. These two concepts are directly dependent upon the *chemical potential energy* (as described in Unit 1), which is the energy contained in the chemical bonds of a chemical species (molecules, ions, or atoms). That is, change in Gibbs free energy is the quantitative equivalent of "chemical potential energy." ΔG depends on the number of moles that are shown in a balanced chemical equation. So ΔG is the *quantitative change in chemical potential energy* during a chemical reaction. Furthermore, you should not be surprised to learn that ΔG_{rxn} is dependent on ΔH_{rxn}, change in enthalpy of a chemical reaction.

There are two other factors that ΔG_{rxn} is dependent upon:

- the temperature of the reaction, T (Kelvin), and
- the change in entropy of the reaction, ΔS_{rxn}.

Change in entropy, ΔS_{rxn}, depends on the number of microstate(s) in the products and compared to the number of microstates in the reactant(s). What is the best way to understand how ΔH_{rxn}, ΔS_{rxn}, and reaction temperature (K) affect ΔG_{rxn}? In other words: Is a particular chemical reaction *spontaneous* at temperature, T? Here is a reaction that produces an explosion:

	Fe_2O_3 (s)	+ Al (s) →	Fe (s)	+ Al_2O_3 (s)	
ΔH_f	–824.0	0	0	–1675.7	kJ/mol
S_f	87.4	28.3	27.78	50.92	J/(mol*K)
ΔG_f	–742.2	0	0	–1582	kJ/mol

Let's do the thermochemical calculations:

$$\Delta H_{rxn} = [(1\text{ mol} * -1675.7\text{ kJ/mol}) + 0] - [1\text{ mol} * -824.2\text{ kJ/mol}) + 0]$$

$$\Delta H_{rxn} = -851.5\text{ kJ/mol}$$

Do you recognize that this reaction is *exothermic*? Please look at Figure 7.5—yes, indeed it does give off heat and light (exoergic rxn)!

Figure 7.5 Thermite reaction
©Albert Russ/Shutterstock.com

Is there a *dispersal of matter* in this reaction that produces energy? Yes! You should be laughing if you are looking at Figure 7.5! It is rather obvious that light and heat are spreading out . . . If you were to observe this reaction from within ~ 10 feet, you would feel its warmth. However, you should be at least 10 m away, 30 ft, for safety sake. What is the energy variable that measures dispersal of matter? *Entropy!* . . . so let's calculate the change in entropy for this reaction:

$$\Delta S_{rxn} = \{\text{entropy of products}) - (\text{entropy of reactants})$$

$\Delta S_{rxn} = [(mol_{Al2O3} * S_{f\ Al2O3}) + (mol_{Fe} * S_{f\ Fe})] - [(mol_{Fe2O3} * S_{f\ Fe2O3}) + (mol_{Al} * S_{f\ Al})]$

$\Delta S_{rxn} = [50.92\ J/_{(mol*K)} + 27.78\ J/_{(mol*K)}] - [87.4\ J/_{(mol*K)} + 28.3\ J/_{(mol*K)}]$

$\Delta S_{rxn} = -37.0\ J/_{K}$

Now that we have both ΔH_{rxn} and ΔS_{rxn}, we can calculate ΔG_{rxn}. Why would we want to do this . . . ? Answer: to determine if it is a spontaneous reaction!

$\Delta G_{rxn} = \Delta H_{rxn} - T\Delta S_{rxn}$

$\Delta G_{rxn} = -839.1\ kJ/mol - [298K * -5.21\ J/(mol*K)/1000] = -839.1 - [-1.55\ kJ]$

$\mathbf{\Delta G_{rxn} = -837.5\ kJ/mol}$

The negative sign means that this reaction is spontaneous at 25°C (298K)!

As shown in Figure 7.6, if someone actually observes this explosive reaction in person, then they see that the iron produced *flows* out of the bottom of the reaction container as molten iron. Iron is in its liquid state because the heat generated produces a temperature that is greater than its melting point—mp = 1565°C. So this thermite reaction gives off an incredible amount of heat! The reactants must be ignited before the reaction can occur; however, once this happens, the reaction generates its own heat (i.e., an exothermic reaction). Thus it is an exothermic process. Furthermore, this heat produces a temperature that melts iron. The molten iron then flows out the bottom of the container. Can you express this fact in terms of the relative number of microstates of the products as compared to those of the reactants? This flow of molten iron is spreading out (i.e., dispersing) the liquid matter (Fe) and energy away from the container. In other words, there are very, very few microstates in the solid reactants, but many, many more microstates in the iron atoms as they flow away from the reaction mixture. Is this dispersion of microstates expressed in our answer? No! This process was <u>not</u> factored into the value of ΔG_{rxn} (−837.5 kJ/mol) that we calculated. Our conclusion is that the <u>actual</u> ΔG_{rxn} is even larger than this calculated value. Yes, "flow" could be factored into the calculated answer, but it would involve chemical engineering principles, which are beyond the scope of this course.

Is there a real-world application for this explosive reaction? Yes—If you are working on a railroad and your job is to "fuse two rails together," then you could use this thermite reaction to connect the two rails. Also, note that the molten iron would cool to form a "continuous metal rail." Consequently, when the train travels on the track, then there are no big bumps between rails—well maybe just a small bumpity-bump.

Figure 7.6 Thermite reaction—Molten iron flows through a hole in the bottom of the reaction container. Source: Author's personal photo.

Sec 7.4 Can an Endothermic Process Produce a Spontaneous Reaction?

As compared to the thermite reaction, when ammonium chloride dissolves in water, it produces ammonium and chloride ions and the solution gets colder. How can this process occur spontaneously without generating heat?

	NH_4Cl (s)	→ NH_4^+ (aq) + Cl^- (aq)	
ΔH_f	-314.55	-299.66	kJ/mol
S_f	94.85	169.9	J/(mol*K)
ΔG_f	-203.08	-210.57	kJ/mol

So we can calculate the change in enthalpy for this process as follows:

$$\Delta H_{rxn} = [1 \text{ mol} * -299.66 \text{ kJ}_{/mol}] - [-314.55 \text{ kJ}_{/mol}]$$
$$\Delta H_{rxn} = 14.89 \text{ kJ}$$

As you can see from this answer, when ammonium chloride dissolves in water, the process is *endothermic* and the solution gets cooler because thermal energy is removed from the water and used in the dissolving process.

The next question is ... When ammonium chloride dissolves—is this process spontaneous? Let's calculate ΔS_{rxn} to find out if matter (ions) and, thus, energy are dispersed:

$$\Delta S_{rxn} = [1\text{mol} * 169.9 \text{ J}/_{(mol*K)}] - [1 \text{ mol} * 94.85 \text{ J}/_{(mol*K)}]$$
$$\Delta S_{rxn} = 75.05 \text{ J}/_{K}$$

Thus, when ammonium chloride is mixed with water, this process produces more microstates as it dissolves. The consequences of this dispersal of ions is that 37.2 g dissolves in 100.0 g of water at 20°C. How do we explain this process in terms of microstates? Answer: Before the crystals of ammonium chloride are dissolved, there are very few microstates because the NH_4Cl ions are aligned (e.g., see Figure 7.1a). After the NH_4^+ and Cl^- ions are dispersed throughout the solution, hence there are many, many microstates. In other words, think about trying to predict where each of these ions are located throughout the solution. There are many, many possibilities—many microstates distributed in the solution.

Is this process *spontaneous*? Let's calculate change in Gibbs free energy:

$$\Delta G_{rxn} = [1\text{mol} * -210.57 \text{ kJ/mol}] - [1\text{mol} * -203.08 \text{ kJ/mol}]$$
$$\mathbf{\Delta G_{rxn} = -7.48 \text{ kJ}}$$

Yes, the dissolving process occurs *spontaneously*. Furthermore, since ΔH_{rxn} is positive, the process is *endothermic* and the temperature of the solution drops as ammonium chloride dissolves in water. Can you think of any real-world applications of this combination? How about a cold pack? It consists of a sealed packet containing NH_4Cl (s) surrounded by water. When an athlete has an inflamed knee (or whatever), then they could use this pack to cool the inflamed tissue. When the cold pack is needed, the athlete breaks the seal and water flows into the solid, he/she shakes the solution, and it gets colder.

Sec 7.5 When Is a Gas-Phase Reaction Spontaneous?

As described in Sec 7.2 (see Figure 7.1), for a given pure substance (element or compound), the gas phase has many more microstates than the liquid phase, which in turn has more microstates than the solid phase. So whether a gas is a reactant or a product makes a big difference in the change in entropy, ΔS_{rxn}, of a chemical reaction. You should be able to estimate the ΔS_{rxn} for a chemical reaction that includes gases <u>without</u> doing any calculations. But how is this done? The answer is to calculate the change in gas moles between the reactants and products. That is, Δn = (sum of gas products) – (sum of gas reactants). Let's look at some examples:

$2\,H_2O\,(l) \rightarrow 2\,H_2\,(g) + O_2\,(g)$	$\Delta n = 3 - 0 = +3$
$Zn\,(s) + 2\,HCl\,(aq) \rightarrow ZnCl_2\,(aq) + H_2\,(g)$	$\Delta n = 1 - 0 = +1$
$6\,CO_2\,(g) + 6\,H_2O\,(l) \rightarrow C_6H_{12}O_6\,(aq) + 6\,O_2\,(g)$	$\Delta n = 6 - 6 = 0$
$N_2\,(g) + 3\,H_2\,(g) \rightarrow 2\,NH_3\,(g)$	$\Delta n = 2 - 4 = -2$

So the ΔS_{rxn} for each of these reactions depends on the *net number of gas moles*, Δn. Next, you need to apply Campbell's rule to these gas phase reactions[1, 2]. Campbell's rule: if one mole of gas is produced ($\Delta n = +1$), then ΔS_{rxn} = 140 J/(mol * K) To estimate ΔS_{rxn} in general, $\Delta S_{rxn} = \Delta n$ * 140 J/(mol * K) Let's look at this estimate for the gas-phase reactions shown above:

$2\,H_2O\,(l) \rightarrow 2\,H_2\,(g) + O_2\,(g)$	$\Delta S_{rxn} = 140\ J/_{(mol\,*\,K)} * 3 = +420\ J/_K$
$Zn\,(s) + 2\,HCl\,(aq) \rightarrow ZnCl_2\,(aq) + H_2\,(g)$	$\Delta S_{rxn} = 140\ J/_{(mol\,*\,K)} * 1 = +140\ J/_K$
$6\,CO_2\,(g) + 6\,H_2O\,(l) \rightarrow C_6H_{12}O_6\,(aq) + 6\,O_2\,(g)$	$\Delta S_{rxn} = 140\ J/_{(mol\,*\,K)} * 0 = 0\ J/_K$
$N_2\,(g) + 3\,H_2\,(g) \rightarrow 2\,NH_3\,(g)$	$\Delta S_{rxn} = 140\ J/_{(mol\,*\,K)} * -2 = -280\ J/_K$

Finally, how do we interpret these results? Any gas phase reaction that <u>produces gas</u> ($\Delta n > 0$) will *increase entropy* ($\Delta S_{rxn} > 0$); whereas, when a <u>gas is consumed</u> ($\Delta n < 0$) the reaction *decreases entropy* ($\Delta S_{rxn} < 0$). How does ΔS_{rxn} relate to the number of microstates? If $\Delta S_{rxn} > 0$, then # microstates are increasing (→ dispersion). If $\Delta S_{rxn} < 0$, then # microstates are decreasing (→ order).

Is the ammonia-producing reaction spontaneous? That is, let's calculate ΔG_{rxn}:

	$N_2\,(g)$ +	$3\,H_2\,(g)$ →	$2\,NH_3\,(g)$		
				<u>Estimation</u>: ΔS_{rxn} = 140 J/(mol * K) * −2 = −280 J/K	
ΔH_f	0	0	−45.90	kJ/mol	ΔH_{rxn} = −91.80 kJ
S_f	191.6	130.7	192.8	J/(mol*K)	
ΔG_f	0	0	−16.37	kJ/mol	

Thus, from the ΔG_f's equation:

$$\Delta G_{rxn} = [2mol * -16.37\ kJ/mol] - 0 = -32.74\ kJ \quad (at\ 298K)$$

Can you predict whether ΔS_{rxn} be positive or negative? Given: $\Delta n = 2(g) - 4(g)$

ΔS_{rxn} = [2mol * 192.8 $J/_{(mol*K)}$] – [(1mol * 191.6 J/(mol*K)) + (3mol *130.7 J/(mol*K)] =

ΔS_{rxn} = −198.1 J/K

Thus, this reaction has a negative ΔS_{rxn} as predicted because $\Delta n < 0$! How about the accuracy of our estimation? The answer is . . . ΔS_{rxn} = −198.1 J/K *as compared to* ΔS_{rxn} = −280 J/K (Δn * 140 J/K). And so this estimation is not that close to the calculated value. However, when you can estimate the sign of ΔS_{rxn} as (+) or (–), then your *chemistry intuition* is good and it is "in the right direction." You should expect to see non-calculating problems similar to this one on your next major exam and on the final examination.

What happens if we heat up this ammonia-producing reaction? Will it become more spontaneous? Say the given temperature is 200˚C (473K). Let's calculate ΔG_{rxn} as follows:

$$\Delta G_{rxn} = \Delta H_{rxn} - T\Delta S_{rxn} = -91.8 \text{ kJ} - [(473K^* - 198.1 \text{ J/mol}^*\text{K})/1000]$$

$$\mathbf{\Delta G_{rxn} = +1.90\ kJ} \qquad \textbf{at 200˚C}$$

Do you notice that the ammonia reaction is *non-spontaneous* at 200°C? This is not a miscalculation. What about a reaction temperature of 100°C?

$$\Delta G_{rxn} = \Delta H_{rxn} - T\Delta S_{rxn} = -91.8 \text{ kJ} - [(373K^* - 198.1 \text{ J/mol}^*\text{K})/1000]$$
$$\mathbf{\Delta G_{rxn} = -17.9\ kJ} \qquad \textbf{at 100°C}$$

Therefore, this reaction switched from exothermic to endothermic at some temperature between 100˚C and 200°C. How can we calculate that temperature? First, assume that the non-standard ΔG_{rxn} is zero, $\Delta G_{rxn} = 0$, and the ammonia system is at true thermodynamic equilibrium at this particular temperature. Thus,

$$0 = \Delta H_{rxn} - T\Delta S_{rxn}$$

Now we solve for T in Kelvin:

$T_{EQ} = \Delta H_{rxn} / \Delta S$ Note that the two negative signs cancel out (–/– = +). First, let's convert kJ in ΔH_{rxn} to Joules:

$$T_{EQ} = (-91.8 * 10^3 \text{ J})/(-198.1 \text{ J/mol}^*\text{K})$$

$$\mathbf{T_{EQ} = 463\ K\ or\ 190°C}$$

Thus, at 190°C the ammonia reaction is at equilibrium—true thermodynamic equilibrium. It is only spontaneous when the reaction temperature is below T_{EQ}!

For another gas-phase reaction—the photosynthesis reaction: Do you feel slightly uncomfortable about estimating ΔS_{rxn} using only Δn (gases)? If "yes," then *good for you*! In photosynthesis, six moles of CO_2 (g) are consumed and six moles of O_2 (g) are produced. Thus, $\Delta n = 0$ but does this mean that you should predict no change in entropy? *No way!* Please note that six moles of water are consumed and no liquid is produced. Thus, you should predict a slightly negative ΔS_{rxn} because carbon dioxide and water are being fused into glucose, $C_6H_{12}O_6$ (➔ order). What is the magnitude of this effect? Change in entropy is less than zero (negative number) but its magnitude is much, much less than $\Delta S_{rxn} = -140$ J/K. Overall, you should have an intuitive idea for the sign of ΔS_{rxn} for gas-phase reactions: Gas-consuming reactions have a *negative* ΔS_{rxn} (➔ order) while gas-producing reactions have a *positive* ΔS_{rxn} (➔ dispersion).

Sec 7.6 What Are the Four Types of Thermodynamic Reactions?

What are the four thermodynamic parameters that determine whether or not a particular chemical reaction is spontaneous? These are as follows: ΔG_{rxn}, ΔH_{rxn}, ΔS_{rxn}, and temperature in Kelvin. How are these four parameters related to each other?

$$\Delta G_{rxn} = \Delta H_{rxn} - T\,\Delta S_{rxn}$$

Which parameter is relatively independent of reaction temperature? Answer: ΔH_{rxn} is constant for a particular reaction regardless of temperature. Which parameter is dependent on reaction temperature? Answer: ΔS_{rxn} . . . because the higher the temperature, the greater influence the factor ΔS_{rxn} has on determining the value of ΔG_{rxn}.

Can we generalize the different effects that these thermodynamic parameters have on the spontaneity of a chemical reaction? Yes. There are four different types of thermodynamic chemical reactions:

Table 7.1 The four types of thermodynamic chemical reactions

Type	ΔH_{rxn}	ΔS_{rxn}	➔	ΔG_{rxn}	Reaction conditions
I	(–) Exo-	& (+)↑entropy	➔	(–) spontaneous at all T	Rxn > spont w/ ↑ T
II	(+) Endo-	& (–)↓entropy	➔	(+) non-spontaneous at all T	Rxn > non-spont w/ ↑ T
III	(–) Exo-	& (–)↓entropy	➔	@ $T < T_{EQ}$ ➔ Spont	Spont only w/ $T < T_{EQ}$
IV	(+) Endo-	& (+)↑entropy	➔	@ $T > T_{EQ}$ ➔ Spont	Spont only w/ $T > T_{EQ}$

Thermodynamic Type I: Example: $2\,O_3\,(g) \rightarrow 3\,O_2\,(g)$ $\Delta n = +1$

- If a reaction is *exothermic*, $\Delta H_{rxn} < 0$, and *disperses* molecules, $\Delta S_{rxn} > 0$, then it is spontaneous at all temperatures, $\Delta G_{rxn} < 0$. When the temperature increases, the reaction becomes more spontaneous.

Thermodynamic Type II: Example: $3\,O_2\,(g) \rightarrow 2\,O_3\,(g)$ $\Delta n = -1$

- If a reaction is *endothermic*, $\Delta H_{rxn} < 0$, and it *increases* molecular order, $\Delta S_{rxn} < 0$, then it is non-spontaneous at all temperatures, $\Delta G_{rxn} > 0$. There is an "*energy cost*" when temperature increases; the reaction becomes more non-spontaneous.

Thermodynamic Type III: Example: $C_3H_8\,(g) + 5\,O_2\,(g) \rightarrow 3\,CO_2\,(g) + 4\,H_2O\,(l)$ $\Delta H_{rxn} = -2220$ kJ

- If a reaction is *exothermic*, $\Delta H_{rxn} < 0$, and it *increases* molecular order, $\Delta S_{rxn} < 0$, then whether or not it is spontaneous depends on the relationship between the reaction temperature and the "equilibrium temperature" T_{EQ}, where $\Delta G_{rxn} = 0$. When the temperature is below T_{EQ}, the reaction is spontaneous. When the temperature is above T_{EQ}, the reaction is non-spontaneous.

- $T > T_{EQ}$ ➔ Non-spontaneous Rxn, $\Delta G > 0$

at T_{EQ} $\Delta G = 0$ true thermodynamic equilibrium

- $T < T_{EQ}$ ➔ Spontaneous Rxn, $\Delta G < 0$

Thermodynamic Type IV: Example: $CaCO_3$ (s) ➔ CaO (s) + CO_2 (g) $\Delta H_{rxn} = 177.8$ kJ/mol

- If a reaction is *endothermic*, $\Delta H_{rxn} > 0$, and *disperses* molecules, $\Delta S_{rxn} > 0$, then whether or not it is spontaneous depends on the relationship between the reaction temperature and the "equilibrium temperature" T_{EQ}, where $\Delta G_{rxn} = 0$. When the temperature is below T_{EQ}, the reaction is non-spontaneous. When the temperature is above T_{EQ}, the reaction is spontaneous.

  - $T > T_{EQ}$ ➔ Spontaneous Rxn, $\Delta G < 0$

  at T_{EQ} $\Delta G = 0$ true thermodynamic equilibrium

  - $T < T_{EQ}$ ➔ Non-spontaneous Rxn, $\Delta G > 0$

Overall, how can you benefit from knowing these four types of thermodynamic reactions? If it is a quantitative problem you are trying to solve, then you have a way to estimate the answer—the sign on ΔG as being (+) or (-). Using this strategy will help you get the correct answer for calculation problems. For qualitative thermodynamic questions, knowing Table 7.1 will be the only way you can reliably get the correct answer. Hint: On the ACS standardized final exam there will be many more qualitative questions than "calculator questions."

Sec 7.7 Summary

In summary, understanding thermodynamics is a rather complex undertaking. However, you can make it much easier if you can develop an intuitive understanding of each thermodynamic parameter, ΔH_{rxn}, ΔS_{rxn}, ΔG_{rxn}, and temperature (Kelvin). In other words:

- ΔH_{rxn},< 0 means exothermic, ΔH_{rxn},> 0 means endothermic process;
- ΔS_{rxn},> 0 means *dispersal* of molecules (more microstates), which increases reaction energy, and tends to increase spontaneity ΔG_{rxn}; on the other hand, $\Delta S_{rxn} < 0$ means *ordering* of molecules (fewer microstates), which decreases reaction energy, and tends to decrease spontaneity ΔG_{rxn};
- Temperature (Kelvin) makes a difference in the energetics of a chemical reaction; does increasing temperature always increase ΔG_{rxn}? No! That is, when ΔS_{rxn} is (–), temperature decreases ΔG_{rxn} because of the negative sign in $-T\Delta S_{rxn}$ and (minus * minus = +);
- ΔG_{rxn}, tells us whether a chemical reaction is spontaneous (negative value) or non-spontaneous (positive value).

How can this knowledge help you in "advanced" courses? In organic chemistry, you need to know whether a given organic reaction will occur. In biochemistry, you will need to know why it is essential to couple a non-spontaneous with a spontaneous reaction. For example, amino acid + amino acid ➔ dipeptide, and this reaction has $\Delta G_{rxn} > 0$ ($\Delta G_{rxn} = +20$ to 30 kJ/mol): that is, it is a non-spontaneous reaction. So what happens in cellular chemistry? This dipeptide non-spontaneous reaction is "*coupled*" with a spontaneous reaction: for example, ATP + H_2O (l) ➔ ADP + P_i (phosphate) $\Delta G_{rxn} = -31$ KJ/mol. This coupling of these two reactions gives a net spontaneous reaction and thus proteins can be built from amino acids. Hopefully, you can use your knowledge of thermodynamics to understand biology and all the other sciences.

Sec 7.8 Technical References

[1] Campbell, J. A. (1985). LeChatlier's Principle, temperature effects, and entropy. *J. Chem. Educ., 62*(3), 231.

[2] Craig, N. C. (2003). Campbell's rule for estimating entropy changes in gas-producing and gas-consuming reactions and related generalizations about entropies and enthalpies. *J. Chem. Educ., 80*(12), 1432.

CPSIA information can be obtained
at www.ICGtesting.com
Printed in the USA
LVOW02s0453231216
518364LV00005B/13/P